Integrated Production and Pest Management

Integrated Production and Pest Management

Editor

Dr. K.R. Nair

GENE TECH

2025

GENE-TECH BOOKS

ISBN 978-81-89729-75-2

Published by: **GENE-TECH BOOKS**®
4762-63/23, Ansari Road, Darya Ganj,
New Delhi - 110 002
Phone: 09899295259, 011-43003222
E-mail: genetechbooks@yahoo.co.in

PRINTED IN INDIA

Preface

Integrated Pest Management (IPM) or Integrated Production and Pest Management (IPPM) is a pest control strategy that uses an array of complementary methods: natural predators and parasites, pest-resistant varieties, cultural practices, biological controls, various physical techniques, and pesticides as a last resort. It is an ecological approach that can significantly reduce or eliminate the use of pesticides. IPM programmes use current, comprehensive information on the life cycles of pests and their interaction with the environment. This information, in combination with available pest control methods, is used to manage pest damage by the most economical means, and with the least possible hazard to people, property, and the environment.

An IPM regime can be quite simple, or sophisticated enough to be a farming system in its own right. The main focus is usually insect pests, but IPM encompasses diseases, weeds, and any other naturally occurring biological crop threat. The IPM approach can be applied to both agricultural and non-agricultural settings, such as the home, garden, and workplace. Reliance on knowledge, experience, observation, and integration of multiple techniques makes IPM a perfect fit for organic farming. For large-scale, chemical-based farms, IPM can reduce human and environmental exposure to hazardous chemicals, and potentially lower overall costs.

This book provides an in-depth understanding of the basic principles of integrated pest management and gives examples of IPM programmes adopted around the world. It highlights the forces at-work in the environment surrounding IPM farmers, clarifies the relevant IPM processes and identifies the influences of IPM on the forces at work, in what direction they act and whether they are likely to be significant or more modest. It is designed for students, teachers, researchers, policy makers, extension functionaries and farmers who are dedicated to IPM approach.

Editor

Contents

1

Principles and Theory of Pest Management

The process of developing agricultural production, water resource management, improvement of health and other activities of mankind, create an environment favourable to the development of organisms competing with man. This organism is designated as a pest but such a designation is not static, since a pest may be damaging and edible at the same time. For example, crickets and grasshoppers are acceptable as food by some people, but can be a curse to rice farmers. A pest problem exists when an organism interferes with human activities or desire, or otherwise competes with man. To rationally minimise or control pest depredations, an holistic approach to suppression is emphasised. The control strategy that has subsequently evolved is called Integrated Pest Management (IPM).

Integrated Pest Management (IPM)

IPM brings together into a workable combination the best strategies of all control methods that apply to a given problem created by the activities of pests. IPM has been defined in various ways but a more scientific definition describes it as, "the practical manipulation of pest populations using sound ecological principles to keep pest populations below a level causing economic injury". The emphasis here is "practical" and "ecological". There are many ways of controlling insect pests but only a few are practical, and fewer are ecologically sound, such that an undesirable citation is created.

Another term we frequently encounter is "intergrated pest control". It is offer used interchangeably with IPM, though in the strictest sense these terms are not identical. Originally, integrated control simply meant modifying chemical control in such a way as to protect the beneficial insects and mites, or integrating chemical and biological control methods. Subsequently the concept was broadened to include all suitable methods that could be used in complementary ways to reduce pest populations and keep them at levels which did not cause economic damage.

This essentially is IPM It includes a variety of options, any one of which may not significantly reduce the Pest population, but the sum total of which will give adequate reduction to prevent economic losses. A modern definition of IPM may be-the use of all available tactics in the design of a programme to manage, but not eradicate pest population so that economic damage and harmful environmental side effects are minimised.

IPM is not a static, unyielding system. It is dynamic, ever-changing, as we develop a better understanding of all factors that affect the system. These factors include climate, alternate host plants, beneficial insects and man's activities. In a narrow sense, IPM means the management of the few important pests generally found on our crops, but consciously or not it must include all insect pests, not only the "key" ones but also the secondary pests, which seldom do any harm. If this were not so, we might suddenly find some of these minor insect pests or even non-pests elevated to the status of serious insect pests because of our failure to consider them in the total scheme.

IPM as a concept is not new, but one that is receiving new emphasis as man looks for better methods to grow and store food for an expanding population, and at the same time preserve his environment. The rationale for using IPM is threefold. First, it can cut production costs mainly by reducing energy inputs. Secondly, IPM can reduce environmental contamination through the judicious use or reduced use of pesticides. And finally, an IPM programme allows for maximum utilisation of cultural practices and natural enemies and physical methods. IPM can be designed to take advantage of the ecological principles governing pest population abundance. This requires a thorough understanding of the role of all the factors responsible for a pest population reaching certain levels at a particular time of the year, or duration of storage.

Basic Elements of IPM

There are four basic elements of IPM: natural control, sampling economic levels, and insect biology and ecology.

The first element of IPM relates to the fullest utilisation of naturally occurring suppressive factors, including any practice by man which will make the total ecosystem less favourable for growth of the insect pest population. Obviously, this requires a thorough understanding of the ecosystem.

The naturally occurring suppressive factors may act directly or indirectly on pest populations. Indirectly, the ecosystem may be managed or altered in such ways as to make the environment more harmful to the pest and thus limit population growth. More directly, protection and the use of beneficial insects may help keep potentially damaging insect pest populations at subeconomic levels In storage, parameters that can be manipulated to control the buildup of a pest population are temperature, relative humidity, moisture content and composition of gases within the storage atmosphere.

The second element is that of using sound economic threshold (ETL) levels as the basis for applying control measures, especially chemical measures. Establishing and using dynamic ETL's provide a basis for delaying the use of insecticides. This permits the maximum utilisation of other control methods, such as the use of beneficial insects.

The use of economic threshold levels implies adequate sampling of all harmful and beneficial insects in the agroecosystem and particularly in any one crop at a given time. The levels found through sampling must then be measured against the economic level established for the crop, the beneficial insects, and the probable population trend of the pest species. The sampler thus becomes a key person in an IPM system.

The fourth element, insect biology and ecology, is essential to the fullest utilisation of the other three elements. Little concerning natural control can be understood without detailed knowledge of the biology and ecology of all the species present. This knowledge is also essential in establishing the role of each species in the system and in determining the amount of damage inflicted by each pest species. Adequate sampling is directly dependent a thorough familiarity of the species involved.

Knowledge of the biology of a certain problem pest will serve as a basis for planning the control strategies and provide operational guidelines for these strategies. It is important to know the relationship between the pest and the crop and the mortality factors, both biotic and abiotic which play a major role in the determination of pest population dynamics. An understanding of the sequential dominance of pests in relation to growth stages could provide the immediate impetus for developing a simple integrated control programme based on minimum pesticide application. By delineating the succession of major pests at different stages of plant growth, the frequency, timing and dosage of insecticide application could be synchronised, hence avoiding pesticide use on a time-wise basis, or the "calender" method. The control programme could then be based on expected pest population at any given growth stage of storage duration.

Determination of Economic Levels

There is a wide variation in the degree to which pests may be tolerated even for the same species in different areas, in different times of the year on different host plants, and in different stages of crop development. Thus, determination of the Economic Injury Level is critical in defining the ultimate aim of any pest management programme, and in delineating the pest population level below which damage is tolerable and above which specific intervention is needed to prevent a pest explosion and to avert significant damage. Stern et al., defined Economic Injury Level as "the lowest pest population that will cause economic damage" while economic threshold level (ETL) of more accurately Control Action Threshlod (CAT) as "the density at which control measures should be applied to prevent an increasing pest population level from reaching the economic injury level". Although the damage or losses at the economic threshold can be tolerated of neglected, it is at this level that every effort should be made to reduce the pest population by various methods.

Determining the Economic Injury Level and Control Action Threshold is generally a complex matter based on detailed operations of pest ecology as it relates to bioclimatology, predation diseases, the effect of host plant resistance and the environmental consequences of applied control interventions. Rabb has suggested the following factors as essential for the determination of the Economic injury Level:

1. Amount of physical damage related to various pest densities.
2. Monetary value and production costs of the crop at various levels of physical damage.
3. Monetary loss associated with various levels of physical damage.
4. Amount of physical damage that can be prevented by the control measure.
5. Monetary value of the portion of the crop that can be saved by the control measure.

From this information, it is possible to determine the level of pest density at which control measures can be applied to save crop equal to, or exceeding the cost of control. The rise and fall of the Control Action Threshold is determined by the importance of the ecosystem, value of the crop, the pest status, and consumer standards. For example, Heliothis zea feeding on cotton boll, has an economic threshold of four larvae per plant and generally requires insecticidal treatment several times yearly. *H. zea* feeding on sweet corn, has an economic threshold approaching zero population in the United State since the consumer will reject sweet corn with any damage or one larva on it.

In special cases, where pests serve as vectors of plant, animal and human diseases, the economic threshold is zero. A single pest attack may cause the death of a valuable tree, a domestic animal or human. A good example is Aedes aegypti which transmits yellow fever. In grain storage, ETL is influenced by consumer attitudes, Export regulations often specify nil-tolerance of any live insects, be it injurious or beneficial

Factors of IPM

Five general types of single component control methods may be used in IPM programmes in stored ecosystems. These are: chemical control, physical and mechanical methods, biological control, host plant resistance and regulatory control.

Chemical control

A variety of insecticides and acaricides have been and are continuously being developed for control of insect pests. However, these chemicals are but one tool and should be used in combination with other tactics in an IPM programme. The total reliance on chemicals has led to a crisis situation However, IPM does not advocate the complete withdrawal of pesticides. That would be impractical. IPM simply demands use of pesticide only when necessary and at rates compatible with other strategies.

Physical and mechanical methods

Physical and mechanical methods are direct or indirect measures that completely eliminate pests, or make the environment unsuitable for their entry, dispersal, survival and reproduction. Physical-mechanical control measures may include environmental manipulation, mechanical barriers, light taps, irradiation, thermal disinfestation, sanitation, etc. Many times, mechanical and physical methods require considerable extra equipment, materials and labour, hence, they may only be economical in certain situations. For field pests, these methods are rather inefficient but in a storage ecosystem, many of the physical techniques are effective and have great potential for use in an IPM system.

Biological control

Biological control may be defined in a narrow sense as "the manipulation of predators or patho. yens to manage the density of an insect population". This definition does not include the naturally occurring control agents, but only parasitoids, predators and pathogens that are purposely manipulated by man. In a broader sense, it includes "the manipulation of other biological facets of the pest life system, such as its reproductive processes (i.e. sterile male technique), its behaviour, the quality of its food and so forth."

There are some constraints to the potential use and success of natural enemies. Predators, parasites and pathogens found amongst the grain will be regarded as contaminants by consumers and grain exporters. Thus, it makes it very difficult to maintain a pest population level that will enable the biological control agents to establish themselves. The use of pheromones is one of the potentially useful biological agents that could be utilised in IPM for monitoring and partially suppressing pest population not only in agricultural fields but in storage ecosystems.

Host-plant resistance

The manipulation of the genetic make up of the host so that it is resistant to pest attack is called host plant resistance. Over the years there have been numerous successes in breeding for resistance to a variety of pests and currently many crops are being selected for this purpose. This approach has not been attempted to any great extent in stored products protection systems Investigations in this field have been few. However, research has provided evidence of the utility of varietal resistance in grain storage. Unless research on varietal resistance to storage pests is integrated with breeding of plants that are resistant to field insect pests and deceases the potential of this tactic in storage IPM is limited.

Regulatory control

Fundamental regulatory control principles involve preventing the entry and establishment of foreign plants and animal pests in a country or-area, and eradicating, containing or suppressing pests already established in limited areas. Under the auspices of various quarantine acts, numerous control measures are implemented in an attempt to exclude potential pests, to prevent spread and to supplement eradication programmes. Ports of entry are the first line of defense against the introduction of new pests. Pests which break through the port of entry are eradicated or contained within limited areas. Quarantine action is used only against insects of economic importance, although it is sometimes necessary to contain insects which are of no economic importance in another country until their behaviour in a new environment can be studied

Trogoderma granarium is a most serious pest of stored commodities and every effort is extended to prevent its spread in international trade. In many countries, imported consignments found to contain *J. granarium* are segregated and immediately fumigated with methyl bromide. Lately, Prostephanus truncates originating from Central America has become a pest of international quarantine importance.

Component integration

Each of the many methods in insect control has its place in IPM. There are many situations where two or more can be used in an integrated programme. Not all methods, however, are suitable for use in every situation. In a storage ecosystem, hygiene and good warehouse management are essential. It provides the framework for other supplementary infestation control methods. An IPM system would therefore supplement sanitation and good warehouse keeping with one or more combination of the following practices:

1. improved harvesting and threshing techniques
2. judicious use of residual insecticides
3. use of fumigants (MeBr; PH_3)
4. use of ambient aeration, and refrigerated aeration
5. atmospheric gas modification (hermetic; CO_2; N_2)
6. thermal disinfestation
7. irradiation techniques
8. insect resistant packaging
9. insect growth regulators: (IGRs: methoprene, hydroprene)
10. biological control (parasites, predators and entomopathogens, pheromones)
11. Use of resistant varieties if possible

12. Storage management (FIFO)
13. Adequate grain cleaning prior to storage storage.
14. Storage design (for pest exclusion, principally for rodont and bird pests)
15. Adequate grain cleaning prior to storage
16. Monitoring, evaluation and inspection of stored commodities, storage structures and their immediate surroundings.

Insect Pest Management for Stored-Products

There is a number of differences for IPM for stored-products compared to field agriculture. There is far greater variety of tactics that could be employed for storage pest management. In both situations, a systems approach is used in order to facilitate monitoring and implementation.

Field Agriculture and Stored-Product Pests Field Production

1. Ecological condition - more complex and dynamic.
2. Sanitation process - difficult to implement.
3. Economic threshold - appropriately applicable.
4. Physical control - often impractical and expensive.
5. Mechanical control - often impractical and expensive.
6. Regulatory control - more complex and require intensive logistical support.
7. Chemical control - considerable impact on the ecosystem.
8. Biological control - field
9. Host - plant resistance - whole plant modification relatively easier.

Stored Product Setting

1. Relatively simple and manageable.
2. Easier and alone could provide complete control.
3. Not generally applicable, food industry often require zero infestation.
4. Most physical factors could be manipulated under storage condition.
5. More feasible by as built - in feature of storage.
6. Detection and implementation generally self-contained.
7. More contained and limited in spatial proliferation.
8. Problem of contamination.
9. Genetic manipulation to render resistance to seed only more complicated.

In both situations three additional elements must be incorporated to the development of IPM: 1) appropriate people, 2) a systems approach, and 3) adequate evaluation. Aside from the individual expertise of the team members it is important that the people involved must develop a good personal and working relationship in the pursuance of the common objectives.

Economic Principles in Pest Management

Entomologists and biologists define pests as those organisms reaching a certain biomass or population level enough to cause physiological damage to crop and eventually impair either the quantity and quality of the produce. The concern used to be concentrated solely on devising control strategies with the aim of increasing the pest-kill efficiency and/or minimising the control's unwanted side effects on the environment. In doing so ,a damage function and a kill-function are determined simultaneously. Perhaps, the ultimate goal of these efforts is to realise the yield potential of crops by eliminating pest hazards which is possible only where there is maximum protection. In years past, maximum protection was closely associated with prophylactic dosages of pesticides.

Economists view pest problems and solutions quite differently. For one, pests are those organisms that destroy what man has produced, compete for resources needed by man in the production of food and fiber, or contaminate the environment in such a way that man finds it unhealthy or unattractive. An economist would not be very particular about the exact levels of pests, damages, and moralities; but would pay more attention to the value of pest damage and the costs of alternative control strategies with or without special consideration to time. Usually, he would not desire for the yield potential of crops because he knows that it does not always pay to spend for maximum protection. In fact, the urge to control pests does occur to him until he finds that the cost of doing so is more than offset by the benefits. Biologists, to a great extent, do not worry about costs in as much as they find delight in quantifying apparent yield increases due to control action.

Almost for two decades now, attempts have been made to reconcile the disciplinary concerns of both the natural and social scientists to benefit the farmers and the public at large. To put their interests together, the US National Academy of Sciences states that: entomologists develop and perfect various methods of insect control and determine their physiological impacts on crops; economists depend on entomologists for information regarding various inputs combined to produce a particular outcome.

To some extent, pest management was conceived to pool together the concerns of people coming from various fields in a most coordinated fashion. Pest management is defined as an ever-changing process of attacking pest problems by applying whatever economically feasible control of combination of controls that produces the best combination of immediate and long-term results in terms of both reduced pest damage and the absence of unwanted side effects. This implies that pest management may utilise a single-component-tactic or it may be accomplished through a complicated integrated pest management system as discussed earlier. Some Terms and Assumptions in Economic Analysis:

In the analysis of crop protection, it is necessary to define what are we putting in and what are we getting from this activity in order to evaluate if such is really worth it.

Input/Cost: The factors of production such as buildings, machines, materials, labour and management skills are called inputs and if valued at their respective prices are called farm costs or expenses. Let x = quantity of input and Px = Price of input. Then Cost = (Px)(X)

Output/Beturn: Through some physical or biological processes, the above inputs are transformed into usable products or output. Similarly, it can be valued by its price to get the return, revenue or benefit. The benefits from pest control is indicated by the increase in yield or added returns.

Let X = yield (increase) and Py = price of corp. Then
Return or Revenue = (Py)(Y)

Pest management is becoming more essential in our modernising agriculture as problems about pests and damages increase with time. As an input it has 5 components: pest control materials, labour and management or knowledge about pests, expected damage, and alternative control methods.

Pesticides: Pesticides used to be the single most important protectors of crops from insects, weeds and pathogens but because of the ever-increasing price and unwanted side effects, nonchemical methods are being tapped. Nevertheless, continued dependence on pesticides in the near future is a reality that pest managers have to reconcile with. Generally, pesticides form a regular feature of the farm cash expenses.

Non-chemical control: This method is likewise called labour control because it relies heavily on a farmer's manual efforts. Non-chemical methods do not necessarily mean that the materials needed for its utilisation are cost-free.

Efficiency: We define the most efficient pest control strategy as one that requires the least combination of labour cash input for a given level of yield increase, their factors held constant. The definition combines the "efficiency" and effectiveness concepts presented by Semple. According to him the efficiency of an insecticide is the minimum dosage required to reduce the population of target pests by a fixed proportion (80%) regardless of pest pressure. Effectiveness, on the other hand, is indicated by the number of surviving insects of the extent of yield damage after the treatment. It is determined not by the "efficiency" of insecticides but by the strength of pest perssure as insecticides by the over-all farm operations. One needs to know the compatibility of his control measure to his pest problems and an effective manager does.

Optimum thresholds: At least, there are two sets of pest population thresholds found in pest management literature, one being a shadow of the other. Studies regarding plant physiology in relation to phytophagous organisms that reveal there is a tolerable level of plant destruction due to pests wherein the crop is still able to recover and remain vigorous. Apparently, economic injury level (EIL), sometimes termed as the biological threshold level (BTL) is the critical pest density level going beyond what is tolerable to crops. With the EIL, biologists justify control actions. For the economists, such could never be unless the cost of control is zero. Economic threshold level (ETL) is the level of pest density at which suppression action is initiated, the control cost at least equals the value of expected damage. Some IPM specialists prefer to use the term control action threshold (CAT) for EIL since it connotes immediately that it is that level control action is necessary. The mathematical expression for ETL below tells the basic relationship between the critical pest density level and economic variables. If we assume d and K as pre-determined (constant) variables then only the cost and crop prices affect the ETL. Note the prohibitive character of cost control action. If prices of pesticides would be very, very high, ETL would also be and control action may not be economically warranted. However, if the crop is of high value or if the market requires high quality products, ETL will be small and the propensity for control action will be high.

$$ETL = c/pdk$$

where

c = cost of control action

r = price of crop

d = damage coefficient

k = kill efficiency

Guidelines to Pest control Decision

A private farmer, faced with a limited cash budget and credit would think twice before he puts in money into an otherwise biologically attractive innovation. He might not be doing detailed accounting of the proposed change but deep in his mind, he calculates and balances the relative benefits and costs of every new undertaking. If the perceived economic net gain is positive and acceptable, then he evaluates the resource requirements and explores ways finance it.

Partial budgeting is perhaps the most popular decision tool employed in the conduct of applied research by the entomologists in an attempt to add economic substance to experimental results. Partial budgets are designed to show, net profit or loss for the farm as a whole, but the net increase in farm income resulting from a minor change in the on-going operations. It contains 4 major items bearing exact information about the changes in monetary terms, namely: added returns, reduced costs, added costs, and reduced returns.

Added returns is the value of the increase in yield and/or the value of quality improvements. Reduced costs is the value of inputs no longer used under the proposed change. If previously the farm is using 6 man-days of labour but the new technology requires only 4 man-days, then the wage that should have been paid to 2 man-days represent a reduction in cost.

Added costs is the value of additional inputs as a result of the proposed change. This is the exact opposite of reduced costs. The classical example is the cost of additional fertilisation, labour and machine service associated with modern farming.

Reduced returns is the value of forgone output. It the proposed change diminishes yield, then it reduces returns. In some instances, the increase in production is shared with the landlord and harvester/thresher so that the value of harvest that went to them is accounted as reduced returns.

The procedure for partial begetting is straightforward. Analysts have just to be keen the details in the change of process. Given data on market prices, he can put values to each change item and determine if such diminishes or adds to returns or costs. After which, he can compute for the net change in income by subtraction the sum of Added costs and Reduced returns from the sum of added returns and reduced costs:

Net gain/loss in income = (Added returns + Reduced Costs) - (Added costs + Reduced returns)

Biological studies on pest control strongly assume that farmers do not control pests; this is the implication of comparing- treated as against the untreated plots. Such standard experimental procedures directly ease out computations involved in the economic analysis but, at the expense of intuitive value of practical content. It does not provide the analyst a grasp of the general and actual problem situation because at this time very few farmers are indifferent to pests. Pesticide consumption growths observed for rice, vegetables and bananas indicate that pest control is very active in our agricultural/farming system. Quite recently, field studies on the control of rice pests use farmers' practice as a separate benchmark for comparison.

It is quite necessary that analysts look for flexibilities in the data set and mold the framework according to the dictates of the local problem situation. In that case, he could come up with a matrix of problem situations and alternatives. For each situation, the dominant alternative may be decided based on at least 2 criteria: (1) magnitude of net change in income (2) resource requirements and accessibility.

Decision-Making Models

In reality, farms are multi-enterprise in nature. There is also substantial interdependency in decision across crops and over time. Uncertainties or lack of information about pests, damage, controls and prices also exist. In other words, the type of control a farmer applies to his main crop influences his control decision for the second crop. Similarly, his control practices in the previous season greatly affect his decision in both present and future cropping. Imperfect knowledge and limited skills concerning pest control decisions and the presence of natural disturbances may render actual decisions "inferior" to that predicted by the partial budget model. And as one tries to develop realistic models, the analytical tools tend to be conceptually complicated and computationally irksome. This leads us to a brief discussion of the various models developed recently.

Whole farm budgets: This is merely an extension of the partial budgets. It recognises interdependent decisions which means that an adoption of technology in one enterprise affects the rest. Thus, the entire farming system is changed. Oftentimes, it will be hard to distinguish the cause from the effect. Taking the entire system as the unit for analysis makes it simpler.

Production Function Analysis: It is basically statistics in form and is widely used in production economic research. A production function tells the average relationship between the farm product and its factors of production. It bears powerful estimates of the precise contribution of each input to the total product. Headley formalised the economic study of pest control and applied this framework in the analysis of pesticide expenditure in US agriculture. He concluded that pesticides are indeed productive inputs placing a $4 contribution to the value of agricultural production per dollar spent on it. He thus predicted an expanding market for agricultural pesticides in the 70's and challenged the technical faculty to develop a less hazardous but comparatively productive substitute.

Reichelderfer criticised the common practice of using pesticides as a proxy for pest levels to explain variability in production. Note that it is the pest and not the control action that directly affects yield. She pointed out that production functions for pest control are unrealistic if they do not express yield as a function of pest levels. Since this practice is necessitated by lack of data, economists are reminded to be careful in the interpretation of results. For one thing, the so-called productivity of pesticides is dependent completely on pest pressure. It is important, therefore, that the economists understand the biological system with which they are working

Benefit-cost analysis

BCA is probably the most comprehensive analytical tool unfortunately, it requires voluminous data and it is usually not workable. BCA treats pest control as an investment or a flow of costs and returns. Though time is the core of B-C analysis wherein discounting methods are used to calibrate the flow of money in order to come up with timeless measures of efficiency the internal rate return (IRR) and the benefit-cost ratio (BCR). The ideal situation is for the IRR to be greater than the market interest rate and for the BCR to be greater than 1. BCA's can be done both at the farm and the community levels. For the latter, we call it extended BCA. This is of great use in analysing pollution problems related to pest control which is quite beyond the concern of farmers and yet it bears tremendous impact to the is shown below (Table 1).

Table 1. Costs of fumigation with methylbromide (MBr) and phosphine generating formulations (PH_3 tablets) for various applications.

	Fumigant	
Situation	*MBr*	*PH_3*
Space fumigation P7.06 per m³ or USD	P200 per 1000 ft³ or P5.30 per m³ or 0.50 per m³	P150 per 1000 ft³ or USD 0.38 Per m³
Block fumigation	P25 per 1000 ft³ or P8.82 per m³ or 0.63 USD per m³	P220 per 1000 ft³ or P7.77 per m³ or 0.55 USD per m³
Container fumigation	P250 per 20-footor or 17.86 USD	P200 per 20-footer or 14.29 USD
	P400 per 40-footer or 28.57 USD	P18 per 40-footer or 27.14 USD

Space fumigation using PhostoxinR for a warehouse of approximately 980,000 cubic feet volume (or 27,750 m³). The quantity of corn stored was 147,000 sacks x 75 kilograms or approximately 11,000 tonnes. The total value of corn at current prices of P2.40 per kilo is P26,460,000.00. The reported loss due to Sitophilus granarius (L.) infestation over a period of three (3) to four (4) months was 10% to 15% of the total weight. In real terms, this is equivalent to 1,102,500 kilograms or P2,646,000.00. The cost of fumigation at the current prices of P150.00 per 1,000 cubic feet, is P148,200.00. Equating this figure to the possible loss if no fumigation is done, there is a saving of P2,497,000.00.

Risk analysis

Finally, the latest add-on to our list views the resurgence of pests as an involuntary risk taken by farmers, pest control actions as means to minimise the risks or variability of yields and profits, and control costs as the premium farmers pay to avoid it. This analysis exerts less pressure on biological data requirements, specifically, pest infestation levels. This type of analysis has the potential of influencing pest control policies particularly crop insurance and pest information services.

Socio-economic and Technical Dimensions if IPM

Pest control has gone through an evolution from the sixties where emphasis was on the use of chemicals as the best technical option against pests, to the eighties where the thrust is on judicious, needbased use of pesticides that integrates biological, cultural, and chemical control methods.. This new pest control concept, referred to as Integrated Pest Management or IPM has been stimulated by the intensification and increasing complexity of crop protection problems, coupled with the associated environmental, financial, and health hazards of heavy chemical use.

Kenmore, defined the meaning of integrated pest management as follows:'....."It's integrated because it tries to reassemble into one unit the best mix of pest controls for a particular field in a particular season or time. The glue that holds these controls together in an integrated system is the natural population control ecology of the community of species that live in the field. Natural forces which include biochemical immunity habitat invulnerability and carnivory keep more than 99% of pests under control 100% of the time......

.....PESTS means any species that eats the crops and

particular crop. Pest may include weeds, rodents, insects, fungi, bacteria, viruses, nematodes, birds and mycoplasms. A species becomes a pest only if it eats too much of that crop or commodity which reduces yield from its agronomic potential......

.....Management means the set of skills ... to be learned and practised.

.....Integrated Pest Management, therefore, is a set of skills that enable farmers, users, of practitioners to recognise which of the few populations out of the many in their fields or warehouses are likely to become pests. IPM then gives these users a large selection of control tactics from which they can choose the best mix of pest controls without necessarily involving use of more pesticides "

In the context of the associated environment and the population dynamics of the pest species, IPM utilises all suitable techniques and methods in as compatible a manner as possible to maintain the pest population at levels below those causing economic injury. This means the rational use of all techniques in suppressing pest population through such measures as biological control, genetic techniques, host plant resistance, and cultural, mechanical and chemical controls. It is an effort to bring together multidisciplinary methodologies in developing agroceosystem management strategies and decision-making tools that are workable and economically feasible, ecologically sound, and socially acceptable.

Technical Dimensions

Pesticides are scarce commodities whose life spans must be maximised to ensure that they are available for continued use, if so desired, within the integrated pest management programmes. Insecticide residues decay over a wide range of temperature and grain moisture contents, thus the need to determine these data to allow for; (a) accurate prediction of the amount of pesticide needed for a specific crop or commodity; (b) avoidance of under- or over-dosing; and (c) possible definition of residue tolerances.

Insect/Pest Biology

A major problem in insect/pest control is the lack of knowledge of pest biology and the factors causing physical loss and quality deterioration of grains in storage. Correct identification/ recognition, of the various pests, their life cycle and behaviour patterns have a considerable bearing on the inspection methods and timing of control application for more effective results. For instance, some egg-laying species like the grain weevils, lesser grain borers and the grain moths, initiate their infestation and damage to the grain by leaving their eggs on the unharvested grain in the field. These eggs then mature and become adults when these infested grains are already in storage causing the most damage.

It is, therefore important to have a knowledge of pest biology because different species require different strategies of control at the various stages of their life cycle. Some storage pests are much less dangerous than others and may not necessarily require immediate pest control. Conversely, others are highly destructive, thus the need for immediate control measures becomes imperative.

Biological control

Another major issue about chemical pest control is the propensity of some pests to develop resistance to pesticides through time. The problem is further aggravated by the limited number of chemical classes of pesticides/insecticide available, and its continuing soaring prices. One possible solution to these problems is the use of biological control measures such as parasites and predators and disease organisms or pathogens such as various subtypes of the bacterium, Bacillus thuringiensis amoebae and sporozoans such as Triboliocystis garnbami, Mattesia trogoderma and other species, *Nosema* spp., *Halics* spp., which are most frequently encountered in dense populations of insects and can cause high mortality. None of these are known to infect man and therefore offer considerable potential for use in biological control.

The use of various botanical insecticides is another cheap pest control method in paddy storage. One good example is the neem plant which was found to inhibit insect population buildup, primarily Sitophilus oryzee and Rhizopertha dominica after three months of storage. The repellant effect of the volatile, biologically active ingredient in neem, such as azadirachtin and melianatrol has been evaluated, the former of which possesses known antifeedant properties against *R. dominica.* Neem leaves or seed in powder form had been previously studied to show long persistency with proven effectiveness against *S. oryzee, R. dominica,* and *Trogoderma granarium* in wheat for nine to twelve months of storage.

Neem can therefore be a cheap alternative to synthetic insecticides, with potential adaptation for large scale use. Extracts from black pepper have bee evaluated in the Philippines and other vegetable oils such as palm oil, bran oil, peanut of,, and corn oil at 5-15 ml/kg have provided insect control for four months on legume seed without affecting seed viability.

Several studies have likewise looked into the issue of varietal resistance as a possible pest control measure. One of these is the study done by Rejesus and Dimaano on the resistance of 20 varieties of milled rice to corn weevil (*S. zeamais*) and red flour beetle (*T. castaneum*). The study showed that IR36 and IR32 are highly resistant to corn weevil and red flour beetle respectively. Varieties such as IR29, IR38, IR42, IR43, IR46 and IR5853 were highly susceptible to *T. castaneum*. These results indicate that insects preferred some varieties for feeding because in some cases, these varieties are suitable for the insects' survival and development. The varieties' chemical composition may likewise explain degree of preference.

Effective loss assessment

According to Hodges, *et.al.*, there had been virtually no scientific evaluation of the methods used in the detection of insect infestation or the interpretation of results that are obtained there from. From what is currently known about distribution and behaviour of insect pests in grain stores, it seems that the levels of infestation are very likely to be inaccurately reported. As a result, expensive control procedures such as insecticide spraying and fumigation may often be applied either too early or too late so that the maximum benefits are not realised.

Farmers usually spray insecticides during crop production when they perceive the need for it even when in most cases, the pest level is below the economic threshold level (ETL), And to think that total area coverage rather, than spot application, is often practised one can imagine the waste of the whole exercise. Furthermore, this so-called economic threshold does not consider the population of beneficial organisms.

It is therefore very important that prior to any rational campaign against insects and pests in stored products there is an updated knowledge on the pest complex occurring in the warehouse, the relative importance of each species, products attacked and nature of damage, and the presence and abundance of the natural enemies of the identified pests. Such information will give an indication on changes in pest situation such as pest resurgence, shifts in pest dominance, and the causes of such altered conditions. It is equally important to obtain reliable and fairly accurate estimates of grain losses caused by insects from which the ETL at which control measures should be applied can be established. This information will aid in developing possible improvements in pest control strategies currently being utilised by applying them at a time that will economically justify their use.

Economic Dimensions

According to Norgaard, there are three interrelated ways by which economics may enter into the design of IPM strategies: (a) the farmer's pest management goals are largely economic; (b) economics being a science of resource allocation can aid in selecting optimal quantities and combinations of pest management inputs; and (c) the adoption of new management practices can be speeded up through an economist's understanding of the incentives underlying farmer's behaviour and the effect of alternative social institutions on these incentives.

Physical and economic environment of affected sectors

The ultimate pest management goal of a rational farmer, trader and/or miller/processor is to maximise profit or to produce enough for his needs. This is done through selective reduction or increase of chemical use, selective use of pesticide with a more narrow action spectrum, and use of new biological and other control methods. This is where the importance of studying the socio-economic component of IPM technology is highlighted wherein farmers, traders and/or millers practice pest management strategies according to their resource capacity, their objectives, and their perception of pest attacks. Their acceptance or rejection of the IPM technology will be partly conditioned by the physical and economic environment within which they operate and this includes the size or scale of their operation, access to capital, infrastructural support and services, etc.

The size or scale of a business operation largely determines the willingness of a potential adaptor or user to accept or use a give postproduction technology. In the case of rice producers, small-sized farms are generally on subsistence level that receptivity to new and/or improved technology such as IPM will be nil or totally lacking. As the size of his operation gets bigger and more of his produce is market-directed, his financial resources improves and he becomes more quantity and quality-conscious in order to avail of any price premium for such products. Thus, he may become more receptive to IPM. The same condition can be said of the traders and processors. As their volume of business expands, the more they tend to improve their facilities and management operation. They begin to be more conscious of the quantity and quality of product they procure and distribute to have a greater share of the rice market.

Risk and uncertainty

Farmers living near substance levels are commonly reported adverse to risk because a reduction in income, even for a relatively short period of time, could seriously affect the wellbeing of the farm family. When analysing why potential adaptors and/or users do not adopt new techniques, the new technique must be measured against their needs as well as their capabilities. Surely these targeted beneficiaries have a right to be skeptical when proposed changes involved some financial outlay as compared to the total income that they might have, for something which is new or different from their regular practice or way of life. Moreover, those "improvements" might be too technical or too cumbersome or complicated for these potential users to fully comprehend and use.

Thus, an understanding of the decision-making processes of farmers can help facilitate the adoption of certain pest management techniques. An understanding of this process and how farmers are influenced by their resource endowments, management skills and knowledge of marker factors can help explain potential constraints in the adoption of technology. Unless these constraints are identified and considered during technology development, the present generation of agricultural research scientists may again fail where their predecessors did; the technology they developed was so efficient and perfect, but it hardly conformed to basic realities and farmers' needs.

Loss assessment

In order to effectively estimate/verify the financial benefits of IPM, there is a need to accurately estimate as much as possible the postproduction losses incurred due to pest infestation, the costs incurred due to this loss and the consequent costs of pesticides used under pure chemical control. This information is very important before any appropriate intervention procedure can be undertaken to elucidate where the major losses are being incurred within the system, the magnitude of such loss and the kind of strategy that is best suited for alleviating or improving the situation, both in terms of weight loss reduction and maintenance of quality. The importance of this exercise may not be as great at the farm level - where volume of grain stored is just sufficient for home consumption - as it is at the commercial and/or government level where big volumes of grains are being stored and handled for market.

Social Dimensions

Traditional value system

According to Castillo, the farmers are claimed to resist technological innovations because of their adherence to deeply rooted traditional patterns. But whether farmers are really to blame for the nonadoption is debatable. Decisions of farmers and/or potential adoptors-users are influenced not only by identifiable techno-economic factors but also sociocultural factors. Central to this problem is the value system. According to Rola and Ocampo, traditional families are characterised by the stability of their value system. The essential feature of such is the complex web of agreed norms of behaviour, a kind of cumulative common wisdom unconsciously accepted by everyone without need of visible proof. This complex web is often developed over hundreds of years and for this reason the society gives value to particular practices.

Decision-making process

Integrated pest management is a new concept of a people-oriented technology and its success or failure will depend on how the clientele group will perceive its benefits vis-a-vis their traditional farming practices on the one hand and possibly how change agents or technocrats relate the technology to its end users on the other hand. IPM involves a lot of management decision-making which calls for serious efforts on the part of the extension agent in his tasks to equip and train the targeted clientele with the necessary technical expertise on which to base his decision.

Because of the developed bias of farmers/users as regards pesticides as an important input in their production, farmers may initially look at IPM as a risk and will therefore hesitate to readily adopt it. Their hesitancy is further compounded by the new activities they have to undertake which may be quite "foreign" and tedious to them. This includes: (a) regular visit/monitoring of pest population; (b) learning about "friendly" pests vs. real ones; (c) deciding whether or not immediate action is needed based on infestation levels they themselves have to determine; and (d) facing the consequence that pesticide application is not a sure guarantee of eliminating pest damage.

IPM offers changes in farming methods, hence, changes also in the farmer's way of life. Its introduction demands that farmers be provided not only with the material farm inputs, but also sets of decision-making tools; what resistant variety to use, when to use pesticides, what pesticides to use, etc. The IPM approach requires regular field monitoring, and farmers are encouraged to define what is the economic threshold level, without which judicious and economical use of pesticides will never be realised... The introduction of IPM, therefore, requires the adaptability of the approach not only to the natural environment and economic conditions but perhaps more importantly, to the attitudes, values, and perception of the small farmer.

Effective extension and delivery system

The mere existence of IPM technology is not enough. The introduction of IPM would be meaningless if it will not be transformed and repackaged to suit the farmer's needs. What is therefore needed is an effective extension mechanism, appropriate diffusion approaches and other information support services on crop protection to make the technology usable by the targeted clientele. These IPM approaches must be developed and modified and made compatible to the social system of small farmers. According to Stuart, the verification of location specificity of IPM technology should be done with the active and wholehearted participation of farmer cooperators as "research partners." It should seek to enable the targeted users or practitioners to experience the research process by comparing the agronomic and economic performance of locally adapted IPM with the user's practices in paired-parcel trials within his own facilities.

Because of the complexity of IPM, not only the farmer or potential adoptor should be involved in the training/ extension process but the potential role of non-traditional IPM audiences such as women, children, and other members of the household or community should also be considered. Studies have shown that women generally take over the responsibility of on-farm storage so there's probably a need to redirect to these non-traditional sectors future training and extension programmes.

Moreover, since women can easily/conveniently relate to women extension workers better than the male, upgrading of these female extension agent's skills in this area need to be looked into. Based on IRRI records of IPM training programme, out of the 171 Asian trainees as of 1987, only 17 were women. Whether or not there is a deliberate attempt to discriminate against women extension agents, the end result seems to indicate that the core of women extension agents has not been given equal chances as their male counteraprts, yet the expectations from them are the same.....

Ecological Dimension

According to Rola, total reliance on chemical control in the past has proven to be a disastrous solution to the long-term problems of insect and other pests. Experiences from many parts of the world show that too much pesticide misuse and overuse create a host of other problems like insecticide resistance, pest resurgence, outbreaks of secondary pests, environmental contamination and increasing food residue levels (Table 2). These latter two problems affect the general population, not just the direct users but the indirect users as well. This is reflected in the available statistics on pesticide poisoning cases shown in Table 3.

Table 2. Some possible side-effects of the production and use of pesticides.

1. Entrance and persistence of pesticides into different compartments of the environment, including the food chain.
2. Occupational hazards associated with chemicals with high biological activity.
3. Unintentional exposure of people as a result of careless spraying operations.
4. Careless transportation, storage, and destruction of pesticides and pesticide containers.
5. Mismanagement of dangerous wastes at the production plants.
6. Ecological effects, including development of resistance in disease vectors.

The table reflects the data gathered from government hospitals in 48 provinces for 1980-83 showing 659 total cases of acute pesticide poisoning in 1980, 633 in 1981, 238 in 1982 and 824 in 1983. There are no available statistics from rural doctors who routinely treat patients for occupation "sickness" during and after the main spraying season, so the above figures are likely underestimated. About 60-71% of the cases were suicidal or non-agricultural related since most of these hospitals are located in the urban areas.

A recent study by Loevinsohn has indicated that in the major rice growing areas of the Philippines, widespread adoption of insecticides by small holder farmers appear to have resulted in an increase in mortality of 27 percent among economically active users as a result of pesticide misuse. This would imply an annual mortality of many tens of thousands in the rice growing areas across Asia whose farmers adopt similar practices. It also indicates that the commonly adopted figure of 10,000 deaths worldwide from accidental and occupational poisoning is understated. That the impact is among economically active men implies that the social or economic impact is greater than the undifferentiated number of deaths would suggest.

Table 3. Pesticide poisoning cases admitted to government hospitals according to chemical grouping, Philippines, 1980-1983.

Chemical Grouping	*1980*	*1981*	*1982*	*1983*	*Total*	*%*
Organosphate	229	246	90	346	911	38.8
Organochlorines	116	114	61	154	445	18.9
Carbamates	82	52	35	129	298	12.7
Pyrethroids	5	5	5	5	20	0.8
Chlorophenoxy compounds	11	70	4	8	93	3.9
Dipyridyle	–	1	–	–	1	–
Rodenticides	19	7	2	11	39	1.6
Fungicides	–	1	–	1	2	–
Herbicides	7	3	6	12	28	1.2
Not Specified	148	114	33	126	421	17.9
Mixtures	32	14	3	26	75	3.1
Other Ag. Chemicals	1	3	2	6	12	0.5
Total	650	633	238	824	2345	100.00

In terms of pesticide residue analysis, the accepted practice have always been to analyse the residues of harvested/milled crop at the market. The local marketing, however, does not allow for the disposal/rejection of produce due to high residue levels. Thus, even if residues found in food exceed the maximum residue limit set by FAO or even our own FPA guidelines, nothing is being done. With regard to our export grain, if they do not pass the quality control of the importing country, then the produce is dumped in our local market. For safety/health reasons, it is important that residue analyses be done at the farm level where, given a certain amount of pesticide being used, one could determine the corresponding residue based on a transformation function. There's likewise a research need that will relate toxicity levels and residue levels.

There is also an urgent need to develop protective covering for farmers/users during pesticide application. Although pesticide companies have already developed protective clothing, they may find difficulty in selling them because of the psychological effects on the farmers/suers. This may further discourage users from using that particular pesticide. Proper and strong extension programmes should be undertaken to make possible farmers education and understanding of the proper use and handling of pesticides.

2

Pest Risk Analysis

Pest risk analysis is set of activities, in which the probability and the severity of impact of a particular pest is assessed, the means of reducing these are evaluated. It helps to ensure that decisions on the control of imports and exports are based on sound science, are related to the risks involved and have minimal impact on trade.

PRA Process

The elements of the PRA process are the identification of the pest(s), determining if the pest(s) meet the characteristics of a quarantine pest in accordance with the International Plant Protection Convention (IPPC) definition, assessing the level of risk and then selecting pest management procedures to deal with this risk. In the case of resource-poor countries, the range of pest management options available is limited, so the number and type of quarantine activities that become the import conditions, is limited.

The PRA process is divided into three phases –

— Pest risk initiation

— Pest risk assessment

— Pest risk management

Phase 1 Initiation

The initiation phase begins with the identification of the pest. In the PRA process for access of a commodity such as a fruit, a seed or a grain, the specific commodity has to be identified botanically and the parts of the plant that form this commodity have to be determined. Once this is completed, a list of pests in your country and also the country of export, can be prepared.

The difficulties are follows:

- — identification of the pests on a particular part of the plant
- — stages of the lifecycle on a particular plant part
- — correct classification of some pests due to changes in taxonomy
- — identification of distribution of a pest to a particular country, or part of a country.

Phase 2 Pest Risk Assessment

To undertake this phase of the process a datasheet is compiled of all the technical information on the biology of the pest. The information for the datasheet is gleaned from as many sources as possible, and the country of export is obliged to provide it on request from the importing country. A standard datasheet format derived from the Australian Quarantine and Inspection Service (AQIS) PRA report is available.

Phase 3 Pest Risk Management

Once the list of quarantine pests is compiled from the long list of potential quarantine pests, those with similar characteristics (biologies) can sometimes be compiled into sub-groups that can be managed in the same manner, i.e. Surface pests can be visually inspected.

The facilities, equipment and the staff available will limit the management options for resource-poor countries. The classification of the pests into one of the three Pacific Plant Protection Organisation (PPPO) categories (critical, non-critical and non-regulated pests) will determine where the treatment will take place.

The type of treatment will depend on availability. In examining the type of treatment to be undertaken it is useful to consider actions that are possible at the various sequential stages of the production, packing and distribution process. The management of pest risk can begin by simply listing a pest in your legislation as being a quarantine pest. It can move to more complex processes, some involving treatments.

Initiation Stage

The aim of the initiation stage is to identify the pest(s) and pathways which are of quarantine concern and should be considered for risk analysis in relation to the identified PRA area.

The PRA process may be initiated as a result of:

— the identification of a pathway that presents a potential pest hazard

— the identification of a pest that may require phytosanitary measure

— the review or revision of phytosanitary policies and priorities.

Identification of a Pathway

The need for a new or revised PRA of a specific pathway may arise in the following situations:

- international trade is initiated in a commodity not previously imported into the country or a commodity from a new area or new country of origin
- new plant species are imported for selection and scientific research purposes
- a pathway other than commodity import is identified.

A list of pests likely to be associated with the pathway may be generated by any combination of official sources, databases, scientific and other literature, or expert consultation. It is preferable to prioritise the listing, based on expert judgement on pest distribution and types of pests. If no potential quarantine pests are identified as likely to follow the pathway, the PRA may stop at this point.

PRA Initiated by the Identification of a Pest

A requirement for a new or revised PRA on a specific pest may arise in the following situations:

- an emergency arises on discovery of an established infestation or an outbreak of a new pest within a PRA area
- an emergency arises on interception of a new pest on an imported commodity
- a new pest risk is identified by scientific research
- a pest is introduced into an area
- a pest is reported to be more damaging in an area other than in its area of origin
- a pest is repeatedly intercepted

— a request is made to import an organism
— an organism is identified as a vector for other pests
— an organism is genetically altered in a way which clearly identifies its potential as a plant pest.

A requirement for a new or revised PRA originating from policy concerns will most frequently arise in the following situations:

— a national decision is taken to review phytosanitary regulations, requirements or operations
— a proposal made by another country or by an international organisation is reviewed
— a new treatment or loss of a treatment system, a new process, or new information impacts on an earlier decision
— a dispute arises on phytosanitary measures
— the phytosanitary situation in a country changes, a new country is created, or political boundaries have changed.

PRA area Identification

The PRA area should be defined as precisely as possible in order to identify the area for which information is needed.

Information gathering

Information gathering is an essential element of all stages of PRA. It is important at the initiation stage in order to clarify the identity of the pest(s), its/their present distribution and association with host plants, commodities, etc. Other information will be gathered as required to reach necessary decisions as the PRA continues. Information for PRA may come from a variety of sources. The provision of official information regarding pest status is an obligation under the IPPC facilitated by official contact points.

A check should also be made as to whether pathways, pests or policies have already been subjected to the PRA process, either nationally or internationally. If a PRA exists, its validity should be checked as circumstances and information may have changed. The possibility of using a PRA from a similar pathway or pest, that may partly or entirely replace the need for a new PRA, should also be investigated.

Pest Risk Assessment

The process for pest risk assessment can be broadly divided into three interrelated steps:

— pest categorisation

— assessment of the probability of introduction and spread

— assessment of potential economic consequences.

In most cases, these steps will be applied sequentially in a PRA but it is not essential to follow a particular sequence. Pest risk assessment needs to be only as complex as is technically justified by the circumstances. This standard allows a specific PRA to be judged against the principles of necessity, minimal impact, transparency, equivalence, risk analysis, managed risk and non-discrimination set out in ISPM.

Categorisation of Pest

At the outset, it may not be clear which pest(s) identified in Stage 1 require a PRA. The categorisation process examines for each pest whether the criteria in the definition for a quarantine pest are satisfied. In the evaluation of a pathway associated with a commodity, a number of individual PRAs may be necessary for the various pests potentially associated with the pathway. The opportunity to eliminate an organism or organisms from consideration before in-depth examination is undertaken is a valuable characteristic of the categorisation process. An advantage of pest categorisation is that it can be done with relatively little information, however information should be sufficient to adequately carry out the categorisation.

Elements of Categorisation

The categorisation of a pest as a quarantine pest includes the following primary elements:

- — identity of the pest
- — presence or absence in the PRA area
- — regulatory status
- — potential for establishment and spread in PRA area
- — potential for economic consequences (including environmental consequences) in the PRA area.

Identity of pest

The identity of the pest should be clearly defined to ensure that the assessment is being performed on a distinct organism, and that biological and other information used in the assessment is relevant to the organism in question. If this is not possible because the causal agent of particular symptoms has not yet been fully identified, then it should have been shown to produce consistent symptoms and to be transmissible.

The taxonomic unit for the pest is generally species. The use of a higher or lower taxonomic level should be supported by scientifically sound rationale. In the case of levels below the species, this should include evidence demonstrating that factors such as differences in virulence, host range or vector relationships are significant enough to affect phytosanitary status. In cases where a vector is involved, the vector may also be considered a pest to the extent that it is associated with the causal organism and is required for transmission of the pest.

Presence or absence in PRA area

The pest should be absent from all or a defined part of the PRA area.

Regulatory status

If the pest is present but not widely distributed in the PRA area, it should be under official control or expected to be under official control in the near future.

Potential for establishment and spread in PRA area

Evidence should be available to support the conclusion that the pest could become established or spread in the PRA area. The PRA area should have ecological/climatic conditions including those in protected conditions suitable for the establishment and spread of the pest and where relevant, host species (or near relatives), alternate hosts and vectors should be present in the PRA area.

Potential for economic consequences in PRA area

There should be clear indications that the pest is likely to have an unacceptable economic impact (including environmental impact) in the PRA area.

Assessing the probability of Pest introduction

Pest introduction is comprised of both entry and establishment. Assessing the probability of introduction requires an analysis of each of the pathways with which a pest may be associated from its origin to its establishment in the PRA area. In a PRA initiated by a specific pathway, the probability of pest entry is evaluated for the pathway in question. The probabilities for pest entry associated with other pathways need to be investigated as well.

For risk analyses that have been initiated for a specific pest, with no particular commodity or pathway under consideration, the potential of all probable pathways should be considered. The assessment of probability of spread is based primarily on biological considerations similar to those for entry and establishment.

Probability of entry of a pest

The probability of entry of a pest depends on the pathways from the exporting country to the destination, and the frequency and quantity of pests associated with them. The higher the number of pathways, the greater the probability of the pest entering the PRA area. Documented pathways for the pest to enter new areas should be noted. Potential pathways, which may not currently exist, should be assessed. Pest interception data may provide evidence of the ability of a pest to be associated with a pathway and to survive in transport or storage.

Identification of pathways

All relevant pathways should be considered. They can be identified principally in relation to the geographical distribution and host range of the pest. Consignments of plants and plant products moving in international trade are the principal pathways of concern and existing patterns of such trade will, to a substantial extent, determine which pathways are relevant. Other pathways such as other types of commodities, packing materials, persons, baggage, mail, conveyances and the exchange of scientific material should be considered where appropriate. Entry by natural means should also be assessed, as natural spread is likely to reduce the effectiveness of phytosanitary measures.

Probability of the pest being associated with the pathway at origin

The probability of the pest being associated, spatially or temporally, with the pathway at origin should be estimated. Factors to consider are:

— prevalence of the pest in the source area

— occurrence of the pest in a life-stage that would be associated with commodities, containers, or conveyances

— volume and frequency of movement along the pathway

— seasonal timing

— pest management, cultural and commercial procedures applied at the place of origin.

Probability of survival during transport or storage

Examples of factors to consider are:

— speed and conditions of transport and duration of the life cycle of the pest in relation to time in transport and storage

— vulnerability of the life-stages during transport or storage

— prevalence of pest likely to be associated with a consignment
— commercial procedures (e.g. refrigeration) applied to consignments in the country of origin, country of destination, or in transport or storage.

Probability of pest surviving existing pest management procedures

Existing pest management procedures applied to consignments against other pests from origin to end-use, should be evaluated for effectiveness against the pest in question. The probability that the pest will go undetected during inspection or survive other existing phytosanitary procedures should be estimated.

Probability of transfer to a suitable host

Factors to consider are:

— dispersal mechanisms, including vectors to allow movement from the pathway to a suitable host
— whether the imported commodity is to be sent to a few or many destination points in the PRA area
— proximity of entry, transit and destination points to suitable hosts
— time of year at which import takes place
— intended use of the commodity (e.g. for planting, processing and consumption)
— risks from by-products and waste.

Some uses are associated with a much higher probability of introduction than others. The probability associated with any growth, processing, or disposal of the commodity in the vicinity of suitable hosts should also be considered.

Probability of establishment

In order to estimate the probability of establishment of a pest, reliable biological information should be obtained from the areas where the pest currently occurs. The situation in the PRA area can then be compared with that in the areas where it currently occurs and expert judgement used to assess the probability of establishment. Case histories concerning comparable pests can be considered. Examples of the factors to consider are:

— availability, quantity and distribution of hosts in the PRA area

— environmental suitability in the PRA area

— potential for adaptation of the pest

— reproductive strategy of the pest

— method of pest survival

— cultural practices and control measures.

In considering probability of establishment, it should be noted that a transient pest may not be able to establish in the PRA area but could still have unacceptable economic consequences.

Availability of suitable hosts, alternate hosts and vectors in the PRA area

Factors to consider are:

— whether hosts and alternate hosts are present and how abundant or widely distributed they may be

— whether hosts and alternate hosts occur within sufficient geographic proximity to allow the pest to complete its life cycle

— whether there are other plant species, which could prove to be suitable hosts in the absence of the usual host species

— whether a vector, if needed for dispersal of the pest, is already present in the PRA area or likely to be introduced
— whether another vector species occurs in the PRA area.

The taxonomic level at which hosts are considered should normally be the "species". The use of higher or lower taxonomic levels should be justified by scientifically sound rationale.

Environmental Suitability

Factors in the environment that are critical to the development of the pest, its host and if applicable its vector, and to their ability to survive periods of climatic stress and complete their life cycles, should be identified. It should be noted that the environment is likely to have different effects on the pest, its host and its vector. This needs to be recognised in determining whether the interaction between these organisms in the area of origin is maintained in the PRA area to the benefit or detriment of the pest. The probability of establishment in a protected environment, e.g. in glasshouses, should also be considered. Climatic modelling systems may be used to compare climatic data on the known distribution of a pest with that in the PRA area.

Cultural practices and control measures

Where applicable, practices employed during the cultivation/production of the host crops should be compared to determine if there are differences in such practices between the PRA area and the origin of the pest that may influence its ability to establish. Pest control programs or natural enemies already in the PRA area which reduce the probability of establishment may be considered. Pests for which control is not feasible should be considered to present a greater risk than those for which treatment is easily accomplished. The availability of suitable methods for eradication should also be considered.

Other characteristics of the pest affecting the probability of establishment are follows:

— *Reproductive strategy of the pests and method of pest survival*—Characteristics, which enable the pest to reproduce effectively in the new environment, such as parthenogenesis/self-crossing, duration of the life cycle, number of generations per year, resting stage etc., should be identified.

— *Genetic adaptability*—Whether the species is polymorphic and the degree to which the pest has demonstrated the ability to adapt to conditions like those in the PRA area should be considered, e.g., host-specific races or races adapted to a wider range of habitats or to new hosts. This genotypic variability facilitates a pest's ability to withstand environmental fluctuations, to adapt to a wider range of habitats, to develop pesticide resistance and to overcome host resistance.

— *Minimum population needed for establishment*—If possible, the threshold population that is required for establishment should be estimated.

Probability of spread after establishment

A pest with a high potential for spread may also have a high potential for establishment, and possibilities for its successful containment and/or eradication are more limited. In order to estimate the probability of spread of the pest, reliable biological information should be obtained from areas where the pest currently occurs. The situation in the PRA area can then be carefully compared with that in the areas where the pest currently occurs and expert judgement used to assess the probability of spread. Case histories concerning comparable pests can usefully be considered. Examples of the factors to consider are:

— suitability of the natural and/or managed environment for natural spread of the pest
— presence of natural barriers
— the potential for movement with commodities or conveyances
— intended use of the commodity
— potential vectors of the pest in the PRA area
— potential natural enemies of the pest in the PRA area.

The information on probability of spread is used to estimate how rapidly a pest's potential economic importance may be expressed within the PRA area. This also has significance if the pest is liable to enter and establish in an area of low potential economic importance and then spread to an area of high potential economic importance. In addition it may be important in the risk management stage when considering the feasibility of containment or eradication of an introduced pest.

The overall probability of introduction should be expressed in terms most suitable for the data, the methods used for analysis, and the intended audience. This may be quantitative or qualitative, since either output is in any case the result of a combination of both quantitative and qualitative information. The probability of introduction may be expressed as a comparison with that obtained from PRAs on other pests.

Assessment of potential economic consequences

Requirements described in this step indicate what information relative to the pest and its potential host plants should be assembled, and suggest levels of economic analysis that may be carried out using that information in order to assess all the effects of the pest, i.e. the potential economic consequences. Wherever appropriate, quantitative data that will provide monetary values should be obtained. Qualitative data may also be used. Consultation with an economist may be useful.

In many instances, detailed analysis of the estimated economic consequences is not necessary if there is sufficient evidence or it is widely agreed that the introduction of a pest will have unacceptable economic consequences. In such cases, risk assessment will primarily focus on the probability of introduction and spread. It will, however, be necessary to examine economic factors in greater detail when the level of economic consequences is in question, or when the level of economic consequences is needed to evaluate the strength of measures used for risk management or in assessing the cost-benefit of exclusion or control.

Pest Effects

In order to estimate the potential economic importance of the pest, information should be obtained from areas where the pest occurs naturally or has been introduced. This information should be compared with the situation in the PRA area. Case histories concerning comparable pests can usefully be considered. The effects considered may be direct or indirect.

Direct Pest Effects

For identification and characterisation of the direct effects of the pest on each potential host in the PRA area, or those effects which are host-specific, the following are examples that could be considered:

— known or potential host plants (in the field, under protected cultivation, or in the wild)

— types, amount and frequency of damage

— crop losses, in yield and quality

— biotic factors (e.g. adaptability and virulence of the pest) affecting damage and losses

— abiotic factors (e.g. climate) affecting damage and losses

— rate of spread rate of reproduction

— control measures (including existing measures), their efficacy and cost

— effect on existing production practices

— environmental effects.

For each of the potential hosts, the total area of the crop and area potentially endangered should be estimated in relation to the elements given above.

Indirect Pest Effects

For identification and characterisation of the indirect effects of the pest in the PRA area, or those effects that are not host-specific, the following are examples that could be considered:

— effects on domestic and export markets, including in particular effects on export market access. The potential consequences for market access which may result if the pest becomes established, should be estimated. This involves considering the extent of any phytosanitary regulations imposed by trading partners

- changes to producer costs or input demands, including control costs
- changes to domestic or foreign consumer demand for a product resulting from quality changes
- environmental and other undesired effects of control measures
- feasibility and cost of eradication or containment
- capacity to act as a vector for other pests
- resources needed for additional research and advice
- social and other effects (e.g. tourism).

Analysis of economic consequences

Time and place factors

Estimations made in the previous section related to a hypothetical situation where the pest is supposed to have been introduced and to be fully expressing its potential economic consequences in the PRA area. In practice, however, economic consequences are expressed with time, and may concern one year, several years or an indeterminate period. Various scenarios should be considered. The total economic consequences over more than one year can be expressed as net present value of annual economic consequences, and an appropriate discount rate selected to calculate net present value.

Other scenarios could concern whether the pest occurs at one, few or many points in the PRA area and the expression of potential economic consequences will depend on the rate and manner of spread in the PRA area. The rate of spread may be envisaged to be slow or rapid; in some cases, it may be supposed that spread can be prevented. Appropriate analysis may be used to estimate potential economic consequences over the period of time when a pest is spreading in the PRA area. In addition, many of the factors or effects considered above could be expected to change over time, with the consequent effects of potential economic consequences. Expert judgement and estimations will be required.

Analysis of Commercial Consequences

As determined above, most of the direct effects of a pest, and some of the indirect effects will be of a commercial nature, or have consequences for an identified market. These effects, which may be positive or negative, should be identified and quantified. The following may usefully be considered:

- — effect of pest-induced changes to producer profits that result from changes in production costs, yields or prices
- — effect of pest-induced changes in quantities demanded or prices paid for commodities by domestic and international consumers. This could include quality changes in products and/or quarantine-related trade restrictions resulting from a pest introduction.

Analytical Techniques

There are analytical techniques which can be used in consultation with experts in economics to make a more detailed analysis of the potential economic effects of a quarantine pest. These should incorporate all of the effects that have been identified. These techniques may include:

— *partial budgeting*: this will be adequate, if the economic effects induced by the action of the pest to producer profits are generally limited to producers and are considered to be relatively minor

— *partial equilibrium*: there is a significant change in producer profits, or if there is a significant change in consumer demand. Partial equilibrium analysis is necessary to measure welfare changes, or the net changes arising from the pest impacts on producers and consumers

— general equilibrium: if the economic changes are significant to a national economy, and could cause changes to factors such as wages, interest rates or exchange rates, then general equilibrium analysis could be used to establish the full range of economic effects.

The use of analytical techniques is often limited by lack of data, by uncertainties in the data, and by the fact that for certain effects only qualitative information can be provided.

Non-commercial and Environmental Consequences

Some of the direct and indirect effects of the introduction of a pest determined in 2.3.1.1 and 2.3.1.2 will be of an economic nature, or affect some type of value, but not have an existing market which can be easily identified. As a result, the effects may not be adequately measured in terms of prices in established product or service markets. Examples include in particular environmental effects and social effects arising from a pest introduction. These impacts could be approximated with an appropriate non-market valuation method.

If quantitative measurement of such consequences is not feasible, qualitative information about the consequences may be provided. An explanation of how this information has been incorporated into decisions should also be provided.

Endangered area

The part of the PRA area where presence of the pest will result in economically important loss should be identified as appropriate. This is needed to define the endangered area.

Degree of uncertainty

Estimation of the probability of introduction of a pest and of its economic consequences involves many uncertainties. In particular, this estimation is an extrapolation from the situation where the pest occurs to the hypothetical situation in the PRA area. It is important to document the areas of uncertainty and the degree of uncertainty in the assessment, and to indicate where expert judgement has been used. This is necessary for transparency and may also be useful for identifying and prioritising research needs.

Pest Risk Management

Pest risk management (in the analytical sense) is the process of identifying ways to react to a perceived risk, evaluating the efficacy of these actions, and identifying the most appropriate options. The uncertainty noted in the assessments of economic consequences and probability of introduction should also be considered and included in the selection of a pest management option.

Level of Risk

The principle of "managed risk" states that: "Because some risk of introduction of a quarantine pest always exists, countries shall agree to a policy of risk management when formulating phytosanitary measures." In implementing this principle, countries should decide what level of risk is acceptable to them. The acceptable level of risk may be expressed in a number of ways, such as:

— reference to existing phytosanitary requirements

— indexed to estimated economic losses

— expressed on a scale of risk tolerance

— compared with the level of risk accepted by other countries.

Technical information required

The decisions to be made in the pest risk management process will be based on the information collected during the preceding stages of PRA. This information will be composed of:

— reasons for initiating the process

— estimation of the probability of introduction to the PRA area

— evaluation of potential economic consequences in the PRA area.

Acceptability of risk

Overall risk is determined by the examination of the outputs of the assessments of the probability of introduction and the economic impact. If the risk is found to be unacceptable, then the first step in risk management is to identify possible phytosanitary measures that will reduce the risk to, or below an acceptable level. Measures are not justified if the risk is already acceptable or must be accepted because it is not manageable. Countries may decide that a low level of monitoring or audit is maintained to ensure that future changes in the pest risk are identified.

Choice of appropriate risk management options

Appropriate measures should be chosen based on their effectiveness in reducing the probability of introduction of the pest. The choice should be based on the following considerations, which include several of the Principles of plant quarantine as related to international trade:

— *Phytosanitary measures shown to be cost-effective and feasible*—The benefit from the use of phytosanitary measures is that the pest will not be introduced and the PRA area will, consequently, not be subjected to the potential economic consequences. The cost-benefit analysis for each of the minimum measures found to provide acceptable security may be estimated. Those measures with an acceptable benefit-to-cost ratio should be considered.

— *Principle of "minimal impact"*—Measures should not be more trade restrictive than necessary. Measures should be applied to the minimum area necessary for the effective protection of the endangered area.

— *Reassessment of previous requirements*—No additional measures should be imposed if existing measures are effective.

— *Principle of "equivalence"*—If different phytosanitary measures with the same effect are identified, they should be accepted as alternatives.

— *Principle of "non-discrimination"*—If the pest under consideration is established in the PRA area but of limited distribution and under official control, the phytosanitary measures in relation to import should not be more stringent than those applied within the PRA area. Likewise, phytosanitary measures should not discriminate between exporting countries of the same phytosanitary status.

The major risk of introduction of plant pests is with imported consignments of plants and plant products, but it is necessary to consider the risk of introduction with other types of pathways. The measures listed below are examples of those that are most commonly applied to traded commodities. They are applied to pathways, usually consignments of a host, from a specific origin. The measures should be as precise as possible as to consignment type and origin so as not to act as barriers to trade by limiting the import of products where this is not justified. Combinations of two or more measures may be needed in order to reduce the risk to an acceptable level. The available measures can be classified into broad categories which relate to the pest status of the pathway in the country of origin. These include measures:

— applied to the consignment

— applied to prevent or reduce original infestation in the crop

— to ensure the area or place of production is free from the pest

— concerning the prohibition of commodities.

Other options may arise in the PRA area, control measures, introduction of a biological control agent, eradication, and containment. Such options should also be evaluated and will apply in particular if the pest is already present but not widely distributed in the PRA area.

Options for consignments

Measures may include any combinations of the following:

- — inspection or testing for freedom from a pest or to a specified pest tolerance; sample size should be adequate to give an acceptable probability of detecting the pest
- — prohibition of parts of the host
- — a pre-entry or post-entry quarantine system—this system could be considered to be the most intensive form of inspection or testing where suitable facilities and resources are available, and may be the only option for certain pests not detectable on entry
- — specified conditions of preparation of the consignment
- — specified treatment of the consignment - such treatments are applied post-harvest and could include chemical, thermal, irradiation or other physical methods
- — restrictions on end use, distribution and periods of entry of the commodity.

Measures may also be applied to restrict the import of consignments of pests.

Options preventing or reducing infestation in the crop

Measures may include:

- — treatment of the crop, field, or place of production
- — restriction of the composition of a consignment so that it is composed of plants belonging to resistant or less susceptible species

— growing plants under specially protected conditions

— harvesting of plants at a certain age or a specified time of year

— production in a certification scheme. An officially monitored plant production scheme usually involves a number of carefully controlled generations, beginning with nuclear stock plants of high health status. It may be specified that the plants be derived from plants within a limited number of generations.

Options ensuring that the area, place or site of production or crop is free from the pest

Measures may include:

— pest-free area - requirements for pest-free area status are described in ISPM Pub. No. 4: *Requirements for the establishment of pest free areas*

— pest-free place of production or pest-free production site - requirements are described in ISPM Pub. No. 10: *Requirements for the establishment of pest free places of production and pest-free production sites*

— inspection of crop to confirm pest freedom.

Options for other types of pathways

For many types of pathways, the measures considered above for plants and plant products to detect the pest in the consignment or to prevent infestation of the consignment, may also be used or adapted. For certain types of pathways, the following factors should be considered:

— Natural spread of a pest includes movement of the pest by flight, wind dispersal, transport by vectors such as insects or birds and natural migration. If the pest is entering the PRA area by natural spread, or is likely to enter in the immediate future, phytosanitary measures may have little effect. Control measures applied in the area of origin could be considered. Similarly, containment or eradication, supported by suppression and surveillance, in the PRA area after entry of the pest could be considered.

— Measures for human travellers and their baggage could include targeted inspections, publicity and fines or incentives. In a few cases, treatments may be possible.

— Contaminated machinery or modes of transport could be subjected to cleaning or disinfestation.

Options within the importing country

Certain measures applied within the importing country may also be used. These could include careful surveillance to try and detect the entry of the pest as early as possible, eradication programmes to eliminate any foci of infestation and/or containment action to limit spread.

Prohibition of commodities

If no satisfactory measure to reduce risk to an acceptable level can be found, the final option may be to prohibit importation of the relevant commodities. This should be viewed as a measure of last resort and should be considered in light of the anticipated efficacy, especially in instances where the incentives for illegal import may be significant.

Phytosanitary Certificates and other Compliance Measures

Risk management includes the consideration of appropriate compliance procedures. The most important of these is export certification. The issuance of phytosanitary certificates provides official assurance that a consignment is "considered to be free from the quarantine pests specified by the importing contracting party and to conform with the current phytosanitary requirements of the importing contracting party." It thus confirms that the specified risk management options have been followed. An additional declaration may be required to indicate that a particular measure has been carried out. Other compliance measures may be used subject to bilateral or multilateral agreement.

Phytosanitary measures

The result of the pest risk management procedure will be either that no measures are identified which are considered appropriate or the selection of one or more management options that have been found to lower the risk associated with the pest(s) to an acceptable level. These management options form the basis of phytosanitary regulations or requirements. The application and maintenance of such regulations is subject to certain obligations, in the case of contracting parties to the IPPC.

The principle of "modification" states: "As conditions change, and as new facts become available, phytosanitary measures shall be modified promptly, either by inclusion of prohibitions, restrictions or requirements necessary for their success, or by removal of those found to be unnecessary".

Thus, the implementation of particular phytosanitary measures should not be considered to be permanent.

After application, the success of the measures in achieving their aim should be determined by monitoring during use. This is often achieved by inspection of the commodity on arrival, noting any interceptions or any entries of the pest to the PRA area. The information supporting the pest risk analysis should be periodically reviewed to ensure that any new information that becomes available does not invalidate the decision taken.

3

Pest Management Biotechnology

Forest Pest Management Biotechnology

Forest pest management is responsible for protecting forestland from insects, diseases, and other factors affecting forest health. The division promotes programmes designed to improve or maintain the health and biodiversity of forest ecosystems. The division's mission is accomplished by evaluating factors affecting the health of trees and forests; utilising integrated pest management techniques to mitigate the effects of factors on the forest ecosystem; and promoting forest health to the public and forest user community.

Interest in protecting forests from insect, disease, weed and vertebrate pests has increased in recent years. This has come about largely because of:

- increased awareness of the destructive capacities of pests;
- the heavy toll they take on supplies of commercial and recreational timber;
- environmental concerns;
- effects on threatened and endangered species; and
- availability of new, specific pesticides.

Forest managers have come to realise that much of the damage caused by pests could have been avoided. With adequate knowledge of pest identification and biology, combined with good forestry management practices, it may be possible to prevent or at least reduce losses due to pests. Trees in a vigorous condition are much better able to withstand damage by pests than trees already under stress.

Pest management should be a part of an overall forest management plan. The need for pest control treatments can often be minimised through wise, long-term forestry practices. The pest control method(s) chosen will depend upon the kind and amount of control necessary, balanced with costs and benefits within legal, environmental and other constraints. The most important principle of pest control is to use a control method only when necessary to prevent unacceptable levels of damage. Even though a pest is present, it may not be necessary to control it. It may cost more to control the pest than to cover damage or losses.

Before making management decisions, managers should evaluate potential pest impacts within the context of the ecosystem in which the organism occurs, as well as the population dynamics of the organism. Will the impact of an organism increase, decrease or maintain its level of damage over time? What part(s) of the tree does the pest affect? How many trees are or will potentially be affected? What will be the longterm impact of these organisms? Does the organism cause permanent or only temporary damage? Insects such as the southern pine beetle damage the cambium layer and introduce fungi that almost always cause tree death. In contrast, many foliage feeding insects cause one-time defoliation from which the tree can recover. Most trees can withstand complete one-time defoliation without significant longterm impact on tree health. However, an organism that has the potential to cause multiple defoliations (such as the gypsy moth) can have a much more detrimental impact on tree and forest health.

Before choosing a control method(s):

— Correctly identify the organism to ensure it is a pest.

— Monitor the pest populations and determine the likelihood of economic damage.

— Review available control methods.

— Know and follow local, state and federal regulations that apply.

— Evaluate the benefits and risks of each available treatment method or combination of methods.

— Determine whether there are any threatened or endangered species in the area to be treated.

— Choose the method(s) that are effective yet will cause the least harm to you, others and the environment.

— Correctly carry out the control practice(s) and keep accurate records.

If other management options do not yield satisfactory results, you may need to apply a pesticide to control an undesirable organism (pest) in the environment. The challenge is to use pesticides in a manner that will cause the least harm to non-target organisms in forests, seed orchards and nurseries, while achieving the desired management goal.

Pesticides are tested and labeled for specific pests, crops and for land-use situations. Use of insecticides, fungicides and herbicides is common in managed seed orchards, forest nurseries, intensive short-rotation plantations, and in Christmas tree production. In general, the most commonly used forest pesticides are herbicides used for site preparation, herbaceous weed control, and in pine release treatments. Insecticides are seldom used in general forest management because of high treatment costs and because some pest insects are highly mobile. Currently, the only disease control treatment common in general forestry field applications is for annosus root rot. Vertebrate animals are sometimes controlled through trapping or hunting, but repellents and poison baits may be employed.

Reduction of Chemical Pesticides Use

The need to reduce or eliminate the use of conventional chemical pesticides, both in agriculture and forestry, has fostered a search for alternative products and strategies that have a much lower impact on human health and the environment. Naturally occurring biological control agents, plant-derived target-specific insecticides, and pest-resistant plants obtained through conventional breeding are among the better-known and most accepted alternatives. However, because effective biopesticides, pest-specific compounds and resistant crops or trees are not always readily available for the management of a given pest, scientists have harnessed the power of biotechnology in an effort to improve the efficacy of existing biocontrol agents, to identify target sites in pest insects and to introduce resistance genes into host plants.

Given the original and noble goal of these efforts - i.e., the development of environmentally friendly pest control tools -, it seems somewhat ironic that the products of pest control biotechnology are the subject of greater scrutiny and fear than their highly toxic chemical predecessors, at least at the time the latter were first being considered for registration. This situation apparently reflects a heightened environmental awareness, both in the general public and among environmental advocacy groups, which has its roots in various publications that started pointing out, in the 1960's, the threats posed by chemical pesticides. Along with this awareness is a perception that - rightly or wrongly - pest control tools derived from biotechnology carry more risk for the environment than conventional products.

Genetically Modified Baculoviruses

Baculoviruses are DNA viruses that occur naturally among populations of many insect species, predominantly lepidopterans. Larvae become infected while feeding on contaminated foliage and typically undergo liquefaction upon death, thereby releasing virus onto the host plant and allowing the infection cycle to start all over again. The principal attraction - from an environmental standpoint - in using these viruses for biopesticide development is their very narrow host range: most baculoviruses have the ability to infect only a few closely-related species, and are totally harmless to other organisms in the treated environment. Some naturally occurring baculoviruses have indeed been registered as pest control products in the USA, including one aimed at the gypsy moth. However, despite their environmental advantages, these viral insecticides have not been commercially successful, largely due to their relatively high cost and slow action.

The slow action of baculoviruses has been the main focus of research efforts aimed at improving their efficacy through genetic engineering. Interestingly, some features of baculoviruses make it possible to introduce foreign genes into their genomes in such a way that they are expressed at very high levels; the proteins produced as a result of expression of the introduced genes are chosen with the intention of causing significant physiological disruption in the host, thereby bringing about feeding cessation and death earlier than in insects infected with the wild-type virus.

Over the past 15 years, many genetically modified baculoviruses have been developed in various laboratories, with some displaying improved efficacy relative to the progenitor wild-type virus. For such candidates, it is customary to seek patent protection, particularly if the type of physiological perturbation caused by the inserted gene is novel. The cost of international patent protection may, however, be prohibitive, given the limitations that the virus' narrow host range imposes on the product's commercial potential - a consideration that may be even more important for viruses aimed at forest pests due to the often cyclical nature of their outbreaks, with long endemic periods during which the viral insecticide is likely to remain on the shelf. Thus, with respect to commercial viability, the host specificity of baculoviruses should be viewed more as an impediment than an advantage.

Research activities aimed at extending baculovirus host range to a greater number of lepidopteran species are, therefore, strategically important, particularly for viruses targeting forest pests. Recent progress has been made in this area through the insertion of genes from another group of viruses into the baculovirus genome; these foreign genes encode proteins that suppress the host caterpillar's immune response, thereby allowing a given baculovirus to replicate in an otherwise non-permissive host. Similarly, the identification of genes, within baculovirus genomes, dictating host specificity may allow scientists to manipulate the host range of a given virus through genetic manipulation.

For registration in Canada and the United States, genetically engineered baculovirus-based insecticides are subject to the same regulations and data requirements as conventional microbial pesticides, but with the following additional considerations: (i) effects of the pesticidal substance produced by the inserted gene, (ii) possible modification of host range, (iii) the stability and persistence of the recombinant virus in the environment, and (iv) the potential for genetic exchange of the foreign insecticidal gene with other naturally occurring microbes.

Small-scale field releases of genetically modified baculoviruses, in agricultural settings, have been approved and carried out in the United Kingdom, in the United States, as well as in Canada. China has also been active, recently, in this area of research, and has established a well-structured body, the "National Biosafety Committee" (NBC), to oversee applications for environmental releases and industrial-scale trials involving genetically modified organisms (GMOs). Some of the field trials conducted so far involved viruses carrying only a marker gene and/or with a gene deletion, with the simple aim of assessing persistence and spread in the environment of a genetically altered virus, whereas others were carried out to assess the field efficacy of a virus modified to express an insect-selective toxin gene.

In the latter case, the modified baculovirus was shown to kill faster, resulting in reduced crop damage relative to controls and wild-type-virus-treated plants; in the former case, persistence and spread of the modified virus were shown to be lower than for the corresponding wild-type microbe, giving support to the claim that a genetically modified virus will be less persistent in the environment than the naturally occurring parent virus. From theoretical considerations alone, a strong case can be made for the lower persistence of recombinant baculoviruses based on the fact that they kill the host faster than the wild-type virus, giving it less time to replicate and, therefore, leaving a smaller secondary inoculum on the foliage.

For a discussion of other environmental and health concerns associated with the use of genetically modified baculoviruses, the reader is referred to McClintock et al. who summarised the data presented by DuPont Agricultural Products to the US Environmental Protection Agency (EPA) in the context of a permit application for a field release of a recombinant baculovirus, and for which the company was able to make a strong case for the overall safety of the genetically modified agent, presenting both experimental data and theoretical arguments.

With respect to the control of forest insect pests, the only baculovirus that has so far been the subject of genetic modifications is the one specific to the spruce budworm, *Choristoneura fumiferana* (CfMNPV), the most destructive pest of conifers in eastern Canada. Among the various genetic modifications attempted, one has proven to be rather promising; in this particular case, the gene inserted into the viral genome does not encode a toxin, but a molt-related protein found in the host *C. fumiferana*. Untimely expression of this gene in infected hosts causes an incomplete molt that results in death of the insect. Small-scale field releases of this virus have not yet been conducted; obtaining permits for such releases could prove somewhat more challenging than for agents aimed at agricultural pests given that containment issues for pesticide applications over tall trees, whether aerial or from the ground, are greater than for low-lying crops.

Thus, none of the recombinant baculoviruses developed since the late 1980's have yet been registered for use anywhere in the world and, as a consequence, none are currently part of operational pest management programmes, either in agriculture or forestry. In addition, given the bad press given to GMOs in recent years, applications for registration of genetically modified baculoviruses are not being given a high priority status by regulatory agencies. Inspite of these considerations, recombinant baculoviruses constitute a pest control avenue that is well worth pursuing, particularly for the management of forest insect pests, where control options are often severely limited and where target specificity may be of major importance considering the greater complexity of the natural forest ecosystem relative to the simpler monoculture-based agro-ecosystems.

As is probably true of many new technologies, once a precedent will have been set for the registration of a genetically modified baculovirus, subsequent registrations will likely be easier given that applicants will then be able to draw upon the similarities within this virus family to build a strong case in favor of their overall safety. In the meantime, the adoption of certain research and development strategies could lead to products that have a shorter transit time in the registration pipeline.

First, it is probably wise to choose the foreign gene for insertion among organisms present in the target pest's ecosystem, as was the case for the aforementioned recombinant virus developed against the spruce budworm. In terms of perceived risks of this technology by the general public, a caterpillar gene involved in molt regulation may seem less threatening than a toxin gene obtained from scorpions, even though the toxin has been shown to be specific to insects and the virus is known to be unable to replicate in mammals.

Thus, the search for new genes originating either in the pest itself or in other organisms in its ecosystem should be encouraged. On the issue of risk assessment, scientists will have to develop effective communication strategies aimed at dealing with objections relative to the deployment of these organisms, particularly those that are not based on sound science, while showing respect for legitimate concerns. Past experience with an experimental field release of a genetically modified baculovirus in Canada suggests that a good communication plan can do much to reduce or eliminate public resistance directed at the use ot such genetically modified viruses.

Notwithstanding the efforts to further enhance the efficacy and increase the host range of baculoviruses through genetic manipulations-both of which should increase their commercial potential-the success of these viruses may ultimately depend on our ability to mass produce them more cheaply and more efficiently. Efforts are being made to replace the current *in vivo* virus production methods with *in vitro* approaches using insect cell lines in large bioreactors. Once these methods have been perfected, it should become possible to produce recombinant baculoviruses of more uniform quality and at lower cost, therefore making them more attractive from a commercial standpoint. For the control of forest insects, there will be cases where commercial viability remains unlikely due to the sporadic nature of the pest and the relative target specificity of the virus; in such instances, it may not be unreasonable for governmental research agencies to consider using public funds for virus production and application, particularly if the forest is on public land.

"Biorational" Insectides

Biorational insecticides may be detined as chemicals that aim at disrupting a physiological function specific to insects or a group of insects. Although the active ingredient of these insecticides is a chemical compound, their insect specificity and mode of action make them far more environmentally friendly than conventional chemical insecticides. Some of these molecules are obtained from natural sources or are synthetic analogs of the natural compounds. In these instances, they are often referred to as "biochemical pesticides". Others, however, may bear little resemblance to naturally occurring substances but are chosen on the basis that they inhibit or antagonise a biochemical function specific to insects.

The identification of such compounds can be greatly aided by biotechnology, whereby the genes encoding insect proteins believed to be suitable targets for inhibition or antagonistic interactions are cloned and used for the development of *in vitro* screening systems. Where the three-dimensional structure of the protein can be determined, computer-assisted design can be used to help identify suitable inhibitors and antagonists in an approach similar to that currently employed for drug discovery.

4

Biology and Control of Storage Pests

Stored foods are particularly prone to rodent attack, with the items concerned varying in different regions. The most common and therefore most vulnerable, are maize, rice, sorghum, millet, barley, oats wheat and cereal products. Much food loss occurs as the result of contamination, supplies being rendered unfit for human consumption by rodents hairs faecal droppings and urine, which are shed liberally as the animals forage nightly for their food.

While rodent attack on stored foods is widespread, estimates of damage are poorly documented. The amount of food lost through direct consumption by rodents is considerable. An average-sized Norway rat eats about 259 of food a day, or the equivalent of 9kg in a year. It was found that small colonies of Norway rats, each with access to sack wheat for 12 to 18 weeks, contaminated 70% of the grain and caused a 4.4% loss in weight. The main monetary loss however, resulted from damage to the sacks. Total losses amounted to 18.2% of the value of the wheat and the sacks.

Three species of worldwide distributions are the most important commensals: the Norway or brown rat the roof rat sometimes called the black rat or shiprat and the house mouse. Their characteristics are considered briefly in here. This paper therefore will discuss the reproduction, ecology and control of commensal rodents which are essentially pests in storage.

Taxonomy and Distribution

The Norway rat (*R. norvgicus)* is essentially a temperate climate spectes. It is more abundant and widely distributed across central Asia, Europe and Northern America. The range of this species continues to expand due to changes in urban environments favourable to its habits and to occasional introductions resulting from the traffic. This species frequently live in and around residences, in cellars stores, warehouses, slaughter houses, docks and sewers. On farms they infest silos, granaries, piggeries, poultry houses, stables, warehoused and dockside structures in port areas.

The roof rat (*R. rattus*) is at home indoor or out, depending on the climate. It is a semi arboreal species, climbing shrubs, vines and trees in habitats ranging from river banks to tropical rain forest. This species inhabits a wide range of buildings in temperate areas, including houses, shops and large foodstore, warehouses, poultry houses, barns, market, restaurants and grain elevators. It also lives in close association with man in many cities and villages in the tropics. This species is more extensively distributed worldwide than is *B. norvegious.* The ancestral home of the roof rat is the southern Asian mainland, Southern China, parts of India, Indonesia and the Philippines but it is also distributed in several countries of the Southern and Northern hemispheres.

The house mouse (*M. musculus*) is distributed in temperate, tropical and semi-desert regions. It habitually infest food storage and other premises in both urban and rural surroundings and it is also found occupying such varied habitats as cold stores, rice, sugarcane and cereal grain fields, garbage dumps, salt marshes, and coal mines. Other rodents develop into commensals where their habits bring them into close contact with man. Most notable of these species are the lesser bandicoot rat, Bandicota bengalensis, and multimammate rat, Mastomys natalensis.

The bandicoot rat is distributed in South and Southeast Asia. While it is mainly a pest of agriculture and the predominant field rat in many parts of India, Bangledesh, Burma and Thailand, it has invaded cities, towns and villages and become the main urban commensal in Bombay, Calcutta, Madras, Dhaka, Rangoon and Bangkok. This species feeds extensively on field crops but also infests food storeswhere it feeds on paddy or wheat, both in warehouses and farmers' houses. The multimammate rat (*M. natalensis*) is regarded as a peridomestic rat in most of Africa where it is found in close association with man. It nests in underground burrows or in dark protected areas when living in human habitations. It is abundant in farm houses as it is in the fields.

Reproductive Activity

Reproductive activity in commensal rats and mice is characterized by early sexual maturity, short gestation period,. post-partum oestrus, breeding throughout much of the year and large litter size. These traits give commensal rats and mice the potential for very rapid population growth and for quick recovery when their numbers are reduced by poisons, traps or other means. Under conditions of optimum climate, surplus food and abundant shelter, commensal rodents population tend to breed throughout the year. These conditions most commonly occur indoors in food stores, warehouses, farm buildings and on ships.

Male animals generally remain in breeding condition throughout the year although the testes may be retracted into the abdominal cavity during periods of cold weather when they might appear to be infertile. Norway rats construct a nest of grass, waste paper, twine or other suitable materials in a separate chamber within a burrow system in the natural spaces within buildings. Likewise, house mice build nest in walls and roof cavities, in stacks of food or cabinets or drawers. Roof rats living outdoors often build in shrubs or trees, constructing them from twigs, leaves, grass and other plant materials. The young for all three species need constant maternal care for at least three weeks after birth.

The important physical elements necessary to sustain commensal rodent populations are food, water and shelter. Their abundance and distribution has a direct bearing upon how many rodents can be supported in a given environment. Commensal rodent populations thrive when all three resources are abundant and close together. The main food sources for commensal rodents are stored food and garbage in urban areas, and in rural areas, field crops, natural vegetation and seeds. Stored food are available in mills, warehouses and godowns, port facilities, food processing plants. Feed bins and corn cribs, waste food and spilled grain are other important food sources. Warehouses containing food stored in bags or in bulk are particularly vulnerable to rodent attack unless they are protected. Improperly stored and handled garbage increase rat problems in the urban environment and it is a major cause of the persistence of rat populations in many cities and towns.

Water , generally available to rodents in urban and rural localities but its supply can be a problem for Norway rats, particularly those living in well designed warehouses or in areas with dry seasons. The Norway rat needs food with a high moisture content or a supply of free water, the lack of which restricts its destribution and spread into new areas. House mice, which utilize metabolic water more effectively, tolerate dry habitats without difficulty and the roof rat can also withstand water deprivation better than the Norway rat.

A rodent population that has been reduced in size recovers slowly at first and then at an increasingly faster rate. As it approaches the capacity of the environment, the primary limitations being food, wateı and shelter, growth slows down and the population tends to level off. In an uncharged environment, the population would be expected to remain at essentially the same level but in practice it tends to fluctuate in size, around the carrying capacity of the environment.

Evidently, a confined colony of rodents competes more fiercely for food, shelter and living space as the population density increases. At very high densities, rats and mice spend abnormal time and energy in aggressive attacks and in the defense of territories. Reproductive success of females can also be seriously lowered as the result of aggressive behaviour.

Controlling Methods

The control of commensal rodents is important to safeguard human health and to prevent economic and other losses. Most control work is directed towards preventing rats and mice from living in and around buildings in both urban and rural areas or eradicating populations that have already become established in them. Thus, varied control methods have been devised to come up with an appropriate strategy for certain types of condition. These include environmental sanitation, physical, chemical, and biological control and other methods.

Environmental sanitaion

Environmental sanitation concerns the orderly management of the stored product. Basically it means good housekeeping, the proper storage and handling of food stuffs and organic waste and elimination of harborage for rodents and other pests. Commensal rodents are opportunistic and readily take advantage of man's misuse of the environment. Poor sanitation generally results in food stuffs and harborage being abundant and easily available to rodents and it enables their rapid establishment. Almost any pile of debris indoors or out, can provide rodents with suitable cover for nesting and breeding. In general, the maintenance of a cleared area, as extensive as possible, can do much to deter rats and mice infesting food stores.

Physical control methods

These have been appropriate measures since they do not basically contaminate the stored products. Some of the approaches are ratproofing buildings where rodents are denied shelter and food. It is the most effective and long-term remedy against rats. Preventing entry of rats and mice is important for successful rodent control. This includes keeping openings to buildings and bins tightly closed when not in use. Heavy wire screening or sheet metal barriers on lower portions of openings will help keep rodents out. However, this is a difficult task since commensal redents like mice can pass through 12 mm apertures and very young rats can enter 14 mm openings.

The method is only ideal where storage activity is less. Trapping is the preferred method of killing or capturing rodents ′in situations where the use of rodenticides is considered undesirable. This technique has to be properly done. Because of the cautious behaviour of rats, they tend to be wary of traps and abundance of alternative foods, makes them difficult to capture or control by this approach. In campings against them, traps should be placed near runs and at other locations where there are clear sings of rat activity. Trapping success can be improved by leaving the traps baited but unset for a few days and using baits of proven acceptability to rodents which include bacon, peanut butter, fresh, smoked of ried fish, and ground meats or bread for Norway rats. Baits that dry out or spoil should be replaced immediately by fresh ones. Traps should be examined daily to remove dead rats, which should be buried or incinerated, and to reset those traps that have sprung.

For economic or other reasons, traps are of little value in controlling large infestations of rats and mice but these are useful in helping to catch the survivors of any poison treatment.

The effectiveness of snap-traps against rodents especially on house mice is largely determined by the sensitivity of the traps, their placement and number of traps used. Traps should be set close to walls and in other areas where active runs are evident.

Electric fences which have been used on occasion, both to exclude and enclose rats in the field, can be effective in protecting stored product but obvious caution should be emphasized for possible shocks or electrocution.

The use of very high frequency sound, ultra sound, has been proposed as a means of preventing rats or mice from freely moving into a building or from one area to another. Field and laboratory studies have generally failed to support these contentions. Reports on the repellency of four commercially available ultrasound generating machines tested against a population of wild Norway rats in a large outdoor enclosure using an uninterrupted beam of sound 0.5 m from where they were accustomed to feed shows initial and partial repellency for a day or so but thereafter the feeding behaviour of the rats was unaffected.

Chemical control

Most measures to control commensal rodents depend on the application of poison incorporated in either bait, dust or water formulation. Rodenticides are usually classified as either chronic ar acute compounds. Of most widespread use and particular importance are the anti-coagulant poison, since these slow-acting compounds are now regarded as first choice rodenticides against commensal rodents in most control operations. Acute rodenticides still have a part to play, but they are principally and most effectively employed in situations demanding a rapid reduction of high-density population.

1. Anticoagulant rodenticides have a physiological action in that they discrupt the mechanism that controls blood-clotting and cause fatal internal hermorrhage to develop. Examples of this rodenticides are warfarin, coumachlor, coumatetralyl, diphacinone, and chorophacinone. The new anticoagulants which are so called "second-generations" are difenacoum, brodifrocoum bromadiolone and flocoumafen.
2. Acute rodenticides. The physiological action of this type is either on the stomach or nervous system.

This is categorized based on hazardinuse such as:

a) Compounds that are highly toxic and extremely hazardous to man and animals, e.g. arsenic trioxide, Flouroacetamide, sodium furoacetate.

b) compounds that are both moderately toxic and hazardous to man and animals, requiring considerable care in use, e.g. ANTU (alphanaphthyl-thiourea), calciferol-zinc phoaphide.

c) compounds or relatively lower toxicity that are least hazardous to man and animals. Some of these compounds are norbormide and red squill.

Attention has been given to the possible use of reproductive inhibitors or chemosterilants for the control of rodents. Field trials in England show some indication of reduction of rat populations due to ceased breeding. However, 6-14 months after treatment, a resumption in breeding was observed. This was attributed to a return to normal fertility of long surviving resident rats and to immigration from nearby populations. The future of suitable chemosterilants in the field of rodent control is uncertain, though this has and advantage over rodenticides in that they pose a less immediate hazard to human beings, pers or domestic animals: The major disadvantage is that rodent populations treated in this manner decline rater slowly. Meanwhile, damage, contamination and disease problems have to be accepted.

Fumigants are also used to kill rodents and their ectoparasites living in accessible areas in buildings, ships or in burrows in the soil. Fumigants most commonly used against rodents are calcium cyanide, methyl bromide, chloropicrin and hydrogen phosphide. These are used mainly indoors for insect control and can also be used against rodents in storage. Experience and skill are required therefore in their application.

Repellants are chemicals that are distasteful to rodents; their use in preventive measures is based on the extreme sensitivity of rodents to certain compounds and odors. A number of compounds have been found to possess repellents. The majority are also odorous or toxic to man however and, besides being difficult to work with, few of them have been found to have a long lasting effect.

Biological control

This includes predation, disease and parasitism and genetic manipulation. There are dangers in the introduction of predators which may become pests themselves. The use of diseases and parasites has temporary effects on the population and rats can easily adopt and become resistant. Most of the potential infections agents like Salmonella also cause disease in man and are therefore unsafe for use.

Genetic mutations through in radiation that induces mutations have been tried on *R.r. mindanesis* and *R. argentiventer*. More research on inducing lethal genes and sterile male strains need to be done before this approach can become practical.

Bird Pests in Grain Stores

Birds belong to the Class Abes of Phylum Chordata. They are the only animals with feathers. Among the vertebrates they are the most highly specialized. They are also the most numerous of vertebrates, with roughly 28,500 species and subspecies as compared with 15,000 for mammals and 20,000 for fishes. As a group they have become adapted to aerial life, although there are a number of exceptions. Adaptive features include light hollow bones, loss of right ovary and oviduct in females of most species; the exceptionally well-developed eyes; the highly specialized lung and air-sac system; the modifications of the forelimbs to form wings; the presence of feathers which although light in weight offer effective resistance to air when the bird is in flight. Like the mammals, birds have a constant body temperature, independent of the environment.

The feet and beaks of birds also exhibit various adaptations to different modes of life and different kinds of food. There are birds whose feet are adapted for swimming, perching, and running. Some feed on insects, others on fruits, seeds. A number of grain-eating birds have become important agricultural pests in many countries. In the Philippines, there are various species of bird pests which feed on such crops as rice and sorghum. Collectively, these birds are known locally as maya, and are also called Philippine weavers.

Major Problem

Many species are attracted by ripening grain crops and to drying and threshing floors at harvest time. Some have developed a close association with the more permanent sources of cereals and cereal products and have become a nuisance in food stores and warehouses.

Grain spilled by careless handling in and around stores attracts birds and regular spillage may lead to establishment of a resident population. Accumulations of grain dust and flour residues around mills and processing plants will also attract birds as well as heavy insect infestation on the surface of stocks. Once attracted to a building, birds quickly learn how to enter. Bird populations inside a store pose some major problems such as:

1. They may settle on top of stacks and peck holes in wooven sacks in order to reach the food inside. This can cause spillage and in extreme cases, collapse of the stack.
2. They will roost and nest inside large buildings unless access is completely restricted. In a warehouse, birds, may damage screens in order to gain entry.
3. Nests in guttering and downpiping can cause blockages and lead to flooding and water damage to commodities.
4. They become hosts for lice and mites which become occasional pests of man when birds nest in buildings.
5. Nesting materials will provide harbourage and breeding sites for several of the stored insect pests.
6. As a result of their activity, bird droppings, feathers and decaying bodies may contaminate the foodstuffs, packaging and handling facilities. Droppings are likely to be infected by food poisoning bacteria Salmonella and also constitute a major source of infection for the diseases known Histoplasmosis, Cryptociccosis and Aspergillosis which are caused by fungal spores.

Classification and Identification

The Philippine weavers belong to the Order Passeriformes (Perching birds) whose common feature is the perching foot. The major bird pests belong to the genera Lonchura (Family Estrildidae) and Passer (Family Ploceldae). Another species, *Padda oryzivora* has also been observed but in limited number and limited distribution. Studies conducted by NAPHIRE-TRED showed that warehouse bird populations are predominantly *Passer montanus* or *Mayang simbahan*. The weavers are relatively tiny birds which build covered nests, often beautifully woven. They are mainly nonmigratory. Species are easily distinguished by their plumage. However, there is no sexual dimorphism among these birds.

Methods of Collection

1. Mist nets—Fine silk or nylon net with 3/4 inch mesh, 7 ft. wide and 18 to 38 ft. long. A taut frame of stout twine crossed by horizontal braces called "shelfatrings" is used in conjunction with the mist net, the net and the shelfstring being tight, while the net is loose. The excess netting is arranged in a loose bag or pocket 3 or 4 inches deep below each shelfstring except the topmost one. A bird striking the net from either side carries the net beyond the shelfstrings and hangs in the pocket of the net. A net properly hung, with 4 shelves, is about 6 ft. high.

   Mist nets are more effective when set on a calm day against a dark background such as tall grasses in the weavers roosting area, trees and buildings. Wind greatly reduces catch. The mist nets should be placed where birds tend to fly back and forth fairly low but where there is a minimum chance of people, and animals crashing into a net. It is best to remove birds from mist nets within a few minutes before they become entangled and die form exposure or injury.

   Mist nets require great care. They are easily tangled or torn by contact with vegetation. A person operating the nets should wear clothing with the fewest buttons, exposed pencils and other things to avoid being entangled to the net in the process of removing the birds. To expedite cleaning of nets, caught birds should be placed in a holding. Between uses, nets are best preserved by removing them from the poles and storing them in plastic or cloth bag. Wet nets should be dried out to avoid fungal attack.

2. "Korag"—This is manually operated net trap composed of two rectangular nets (1" mesh, light net) each measuring 1 1/2 m x 2 m. The widths are framed by light poles and the lengths by nylon stings. The nets are set flat on the ground, parallel to each other 3 meters apart such that when activated, they close in perfectly. The nets are hung at the ends and are flipped over the birds by a jerk on the pullcord by the operator.

   The operator must be hidden from the birds by a blind. Decoy birds are needed to attract the birds. The "Korag" can be used where mist nets are ineffective such as in newly harvested rice fields and open grasslands. The trap area should be level and cleared of vegetation and other debris.

3. Modified Australian crow trap—this trap is a large cage made of mesh wire with wooden frames measuring 1 1/2 m wide x 2 m lŏng x 2 m high. The midsection of the truncated V-shaped cage top is provided with holes or slots through which birds can enter. This trap is self-operating and so must contain food, water and some of the captured birds as a decoy.

4. "Kaliked"— this is a funnel-shaped bird trap made of bamboo sticks or coconut midribs used to catch

brooding birds. The trap is fitted into the mouth of the nest and another entrance is opened for the opposite end of the nest. The bird is trapped as it seeks its way out after incubating the eggs. A trap could only catch one bird.

Biology and Behaviour

Not much is known about the Philippine weavers. Field observations reveal that the main diet of these birds consists of rice and Echinchloa seeds. They are also observed to feed on corn tassel, sorghum and certain algae. Feeding begins just after dawn until about 10:00 A.M. and at about 3:00 P.M. to dusk.

The sparrows are primarily seed-eaters. In cage and field experiments, the birds were observed to consume 30% of their body weight per day. The same study also revealed that birds infesting private stores are 15% heavier than those found at NFA warehouses. Apparently, the "openess" or loose construction of private warehouses provide easy access by birds to feed inside. On the other hand, NFA-constructed warehouses partially excluded populations through its better design and provision for bird proofing such as screens. Another reason for the observed discrepancy in weight is the presence of hog and poultry pens around the warehouse. Feeds are also available in these pens and birds obviously feed on them. Results further show that grain composed 91-97% of the gizzard contents of birds collected from private and NFA stores. Other components are weed seeds, grasses and stones.

Birds are abundant in the ricefields especially around harvest, in towns and villages. They roost on pill, starapple, ipil-ipil and bamboo trees as well as houses including warehouses. Their populations follow a definite cropping pattern in an area, that is, it increases when the grains are formed and ripened. Aside from cereals, they also subsist on weed seeds, especially *Echnochloa* spp. and some algae. Philippine weavers are gregarious. Different species are often together is feeding, roosting and even in nesting activities.

The breeding season of these birds may start as early as February with peak of nest building and egg laying in April and May. The breeding season may extend until October. The male and female birds work together building nests in stands of tall grasses, palms and citrus. Mating was also observed to occur in the process. In a warehouse, they build nests on walls, between window slots, crevices of walls, beneath the roofs, along gutters and on the sliding mechanism of doors.

An average of 6 small eggs (5-8) are laid in the nest at one day interval. The eggs are oval measuring 16 mm (greatest length) by 11 mm (greatest diameter) and white or pale ochraceous-salmon in color. Eggs are hatched after a 10-day incubation period.

Newly hatched birds are naked with eyes closed. On the 7th day after hatching, chirping becomes audible while on the 8 th day, eyes start to open and primary feathers appear. The body and feathers develop further and the birds are ready to fly when they are 17-19, days old.

Population and Loss Estimation

There are four forces affecting the size of population:

1) nasality (births)
2) mortality (deaths)
3) immigration movement
4) emigration

In combined form:

$$Ni + 1 + Ni + (B-D) + I - E$$

where:

Ni + 1 = population at time ti + 1 (final)
Ni = population at time ti (initial)
B = births
D = deaths
I = immigrants
E = emigrants

The population growth of birds is characterized by a logistic growth curve:

Characteristics of curve

1. Slow rate of increase to start with
2. Faster rate of increase at greater population size
3. Flatters off to an asymptote

Thus, a pair of birds can give rise to 1,720 individuals in a span of one year, potentially capable of consuming 927 kg and causing a variety of losses through reduced grain quality due to contamination, damage to structures, etc.

Types of population estimations

1. relative density-abundant, common, rare
2. absolute density-number per area

1. *Measurement of relative density*

   – indices of abundance such as droppings, nests and amount of food consumed.

Rice hulls that remain after the birds feeding on paddy may be collected. Since sweeping may still contain paddy. These should be removed and the rice hulls weighed.

From this, the population and/or loss may thus be calculated:

$$\text{weight of brown rice consumed of loss per day (kg.)} = \frac{\text{ave. wt. ricehulls collection}}{0.23}$$

weight of palay loss per day (kg) = wt. of rice hulls collected/day + wt. of brown rice consumed per

$$\text{population} = \frac{\text{weight of brown rice / loss / day(kg)}}{\text{0.006 kg 5 bird / day} \times 0.91}$$

$$= \frac{\text{weight of brown rice loos / day}}{\text{.0055 kg / bird / day}}$$

Egg collection and pest destruction

Regular attention to the destruction and removal of eggs and nests will help to reduce the resident bird population. This needs to be carried out regularly because birds are very reluctant to change nesting sites, even after disturbance, and they tend to return again to the same place.

Traps

A smaller portable version of the modified Australian crow trap (1 m long x .05 m wide x .75 m high mesh wire cage with a V-shaped top with slots trough which birds can enter) has been tried and was partially successful. This is a very time-consuming method because the traps need to be attended to twice daily. Cages should be baited with an attractive food material.

Foot Stickers (glue)

Slow drying plastic jellies can bé applied to sedges where birds roost. These give the perching bird an insecure sensation. Grease can also be used for this purpose. The jellies are applied to horizontal surfaces only, using a caulking gun. The jelly remains effective until clogged with dust residues.

Shooting

This is of limited value but may help to prevent bird roosting in certain areas.

Noise

Noise, and other forms of explosion are likewise of limited value.

Ultra-sonic disturbances and recorded bird call (warning cries and distress call)

These are not reliably effective indoors because the sounds are reflected off in internal surfaces. This causes a modified signal to which birds eventually become accustomed.

In general, physical methods are capital-and labour-intensive especially if birds stay above the piles.

Biological methods

In the field, bigger predatory birds and other animals feed on the weavers. Rodents and cats may also be considered as predators inside stores. Even people catch them at their roosting places at night for food or for trade.

Chemical methods (poison baiting)

These methods should be considered as a last resort if all

appropriate alternative methods have failed. Baiting involves the use of avicides which are potentially hazardous materials and these should be introduced only where safety and supervision can be kept under control. The use of bird poison of pesticidal products intended for other purpose is not recommended. It is unnecessary and dangerous. The food bait for the poison has to be very carefully chosen and it must suit the species of bird. For example, pigeons prefer peas, wheat and cracked or whole corn; sparrows prefer small grains such as palay. All these birds will take bread or cake very readily. The bait offered must, however, be, more attractive that available foods. Pre-baiting 34 days increases the chances of success. For pre-baiting, the same sort of food to be used for the poison bait it is put out in the same place each day.

Repellents like methiocarb discourage birds from further feeding on the seeds. Alpha-chloralose is a material which at low doses has the effect of stupefying birds without killing them. Thus, it is possible to immobilize, collect and remove troublesome burds. They can be released at a safe distance from the warehouse, or destroyed it they are not protected species. As advantage of this it that if non-target species happen to consume the bait, they too will be only temporarily affected. When used at a higher dosage rate, alpha-choralose acts as acute poison and there is no recovery. The concentration of poison to be incorporated in the bait to produce either stupefaction or mortality requires careful determination under local conditions.

Neurological frightening agents such as 4-aminopyridine prepared in grain baits have been used successfully. The material (4-AP) causes birds to emit distress cries and perform erratic flight displays. The bird showing these symptoms will die but it may be sufficient to kill only few birds because the affected birds serve to scare away the rest of the flock. Passer montanus is more susceptible to 4-AP than methiocarb. This species also gave the most audible and greatest number of distress calls in response to 4AP. The use of 4-AP against sparrows in paddy fields and storage warehouses shows promise.

Chemosterilants are used to sterilize a pest population of birds thereby controlling its breeding. The method, however, has disadvantages. The less-toxic chemosteritants are too short-term in action whereas those with more persistent effects are much more toxic. Furthermore, since there is no immediate reduction in population, birds continue to cause a nuisance for some time and may even propagate further because additional birds are attracted to the site by the bait.

None of these agents should be placed where poultry and other animals have access to them. If affected birds are eaten by cats of dogs, they cause secondary poisoning of the predator. For all chemical control measures, a high degree of supervision is essential. This must include regular and frequent collection of dead birds (more than once a day), the burying or destruction of the dead birds, the final removal of all treated bait and the proper disposal of unused bait.

Post-harvest Microbial Infection of Cereal Grain

Spoilage of grains after harvest is due to the interaction of prevailing environmental conditions and the various organisms that attack and contaminate the grains. The environment factors are temperature, moisture, light, gases, and chemicals that may be present. The organisms responsible for grain deterioration are rodents, insects, mites and microorganisms. Changes in the non-living environment provide conditions that may either stimulate or inhibit the activity of the living organisms and thus enhance or prevent spoilage as the case may be.

Grain spoilage is brought about by any of the following microorganisms:

1. Bacteria, which morphologically appear as a single cell or groups of single cells that may be spherical or rod-shaped. They reproduce by binary fission or simple cell division. They need a natural opening or wound to gain entrance to the host as they cannot penetrate the intact grain.
2. Actinomycetes, which are closely related to the bacteria but with elongated cells and form branches. These are usual saprophytic. Many form antibiotics which inhibit of kill other microorganisms in a storage bin.
3. Yeasts, generally appear as single cells and multiply by budding. They predominate in sealed silos where the oxygen supply is low and the moisture content is high. They may impart a fermentation or yeasty odor to the grains.
4. Molds, fungi that have filamentous vegetative structures called mycelium. Reproduction is complicated. They form sexual spores (after the fusion of two compatible cells) as well as a sexual spores (without previous fusion of cells). The spore is more resistant than the mycelium to adverse conditions. When a spore lands on a substrate (such as grain of rice) it germinates and sends out thread-like structures (mycelium) which grow and branch and soon colonize the entire grain, and later produce more spores to complete the cycle. Spores

are continuously present in the soil and air, and are spread by air currents, insects and other agents.

Deleterious Effects of Microbial Contamination

1. *Loss of seed viability:* The germ or embryo is a preferred site of microbial attack because it is delicate, thin-walled and it contains the nutrients needed for microbial growth. Once the scutellum is affected, the food supply for the germinating plant is destroyed.
2. *Altered nutritional value:* The formation of free fatty acids, lowered protein digestibility and vitamin changes have been associated with deteriorative changes in grains brought about by microorganisms.
3. *Health hazards:* Grain dust contains fungal spores and bacterial cells along with a number of substances that cause a variety of symptoms and ailments such as respiratory disorders, eye irritations, skin itchiness, allergies, etc. Aspergillus fumigates can cause lung diseases in man when spores are inhaled. A. *flavus* produces aflatoxin which is carcinogenic and poisonous to several animals, possibly including man.
4. Spoiled grain causes a reduction in the milling yield and milling quality. Such grains are very friable.
5. Contaminated grains are often unpalatable with a musty or sour flavor.
6. Microbial infection may cause of granular products.
7. Molds may cause deterioration of packaging and sealing materials which often result in grain spillage.

8. Microorganisms cause heating of the grain. As the microorganisms respire, they emit heat and moisture which usually make the micro-environment more favourable for further microbial growth. The formation of hot spots in bulk-stored grains is mainly due to the metabolic heat from microbial respiration.
9. Certain microorganisms bring about grain discolorations such as blackening, reddening and yellowing. Yellow grain has become a major problem in the humid tropics. The percentage of yellow grains is high in wet palay that had been left unthreshed for a few days. The high moisture in the unthreshed palay allows for rapid microbial growth. This results in the release of metabolic heat from the respiring molds, causing an increase in the grain temperature in the unthreshed pile. The high temperatures, along with the various secondary metabolites released by the microorganisms, are the likely causes of grain yellowing.

Factors Affecting Microbial Growth and Development

1. *Temperature:* Temperatures from 20 to 40°C favor the growth of most microorganisms. However, some may grow at as low at -90°C (psychrophilic) and at as high as 80°C (thermophilic). Temperatures below 10°C generally inhibit microbial growth but is not lethal in most cases. Thus, a return to favourable temperature for growth, after a period of cold storage, often results in resumption of luxuriant growth.
2. *Moisture content of grains:* Microorganisms are typically moisture loving and a certain amount of moisture in the substrate is required for growth.

The amount of moisture determines the type of organism that is capable of growing in the grain, and the rate at which it grows. Among the various microorganisms, bacteria and actinomycetes are the most hydrophilic followed by the yeasts, and finally the molds which are more droughtresistant. The microorganisms are inhibited at moisture contents in equilibrium with below 70% relative humidity. Fungal spores are not destroyed by dry conditions; their germination and growth are merely inhibited until the moisture in the grain rises to a favourable level.

3. *Length of time the grain is stored and prevailing storage conditions:* The expected storage life of the grain is inversely related to temperature and moisture content. The lower the storage temperature and moisture content, the longer the storage life. Cleanliness in the storage bin and warehouse is a prime requisite in prolonging grain storage life.

4. *General conditions of the grain:* Microorganisms can easily penetrate and initiate infection in damaged or cracked grains. Grain that had been previously invaded by bacteria and fungi is more easily prone to damage than grain that has had no contamination.

5. *Presence of insects and mites:* Insect infestation usually accompanies mold invasion. Insects and mites serve as vehicles for the spread of molds and other microorganisms. Insects are attracted to grains that harbour molds because the insects often feed on the molds as well as on the grain.

6. *Oxygen:* The biochemical oxidation of food materials to obtain energy for life processes is carried out by some microorganisms only in the

presence of oxygen in amounts close to those present in air at atmospheric pressure. Other microorganisms can do with less oxygen and still others grow without oxygen. The molds are strongly aerobic so they grow profusely on the surface of the commodity where athmospheric oxygen is in ample amounts. Generally, yeasts and bacteria can do with less oxygen than the molds (microgerophilic).

7. *Light:* The ultra-violet portion of the spectrum is lethal to microorganisms, but due to its low penetrating power, only those organisms present on the surface of the commodity are killed. Many microorganisms grow best in the dark although some require light for sporulation.

Controlling Microbial Spoilage of Grains

Since microorganisms are present everywhere, there is no practical means of totally eliminating them. However, they may be kept under control by providing conditions that will inhibit growth. The following are recommended:

1. Dry harvested crops promptly to a safe moisture level (13-14% for maize and rice). This is by far the most practical means of preventing microbial spoilage. Prevention is the key control measure because once invasion occurs, with the subsequent deteriorative changes such as aflatoxin contamination, practically nothing can be done to improve the quality of the grain or to remove the toxin. After the grain has been dried to the desired moisture level, rewetting or moisture absorption should be prevented.
2. Provide aeration in bulk-stored grain to prevent moisture condensation which is highly favourable

for microbial growth. Aeration cools the mid-portion of the stored bulk and lowers temperature uniformly throughout the stored grain. This prevents moisture migration. Aeration should, however, not be done at night or when the relative humidity is high because the grains may take up moisture. Increased moisture content will then favor the growth of microorganisms.

3. Prevent insect and mite infestation. In addition to the direct damage they cause, these pests disseminate fungal spores and activate the growth of molds.

4. Low temperature storage (at 10°C or lower) inhibits microbial growth but this method may be expensive and impractical for bulk stored cereals.

5. Storage in air-tight structures has been studied for the control of mold growth. However, bacteria and yeasts may develop under these conditions which often results in off-odors to the commodity. This renders the grain unfit for human consumption because the objectionable odor may be carried over to the processed products.

6. Fungicidal treatment to keep the molds in chect could be expensive and there is the added problem of toxic residues that may render the food or feed unsafe for consumption. Grains intended for seeding purposes may be chemically treated. Fungicides may also be difficult to apply and may adversely affect the processing of grain.

Mold growth can be controlled by 1-2% ammonia in high moisture corn (22-28% moisture content) for 6 months. However, bacterial growth may continue after an initial drop in their number. Calcium propionate or sodium propionate at 5000 ppm can lengthen the storage life of rough rice.

Drying, coupled with proper ventilation or aeration appears so far to be the most effective, the safest, and the most practical means of maintaining the quality of grains. Periodic sampling and testing for moisture content, temperature and presence of microorganisms will allow the detection of potential trouble before anything serious occurs.

Commodities that yield toxin-forming fungi may not necessarily contain toxin(s) because morphologically identical strains vary in their ability to form a toxin and the fungi observed may just be superficially present on the commodity. On the other hand, grains that appear free of molds may be contaminated with a toxin if the toxin-forming mold had been eliminated through desiccation, high temperature or by some other means.

5

Methods for Storage Insect Pests

The inspection of storage facilities and the stored food commodities that they contain, is of paramount importance in preserving grain for human consumption. Stored grain is not an inert substance, but a respiring, living organic entity. It deteriorates during storage, either quantitatively by the amount of dry matter lost (DML), or qualitatively by discoloration, mould contamination, sprouting and caking, etc., promulgated by the activity of microorganisms, invertebrate and vertebarate pests. Further weight losses are incurred during the handling, transporting and processing of stored food commodities.

Accurate information on all these forms of loss can only be obtained by thorough and regular inspection and sampling procedures. This is imperative in formulating sound management decisions involving the adoption of any remedial action against these biodeteriorating agents or the disposal of grain with due cognisance to the condition of both the commodity or the storage facility. Regular inspection helps maintain a storage environment which is conducive to good grain quality by monitoring any significant buildup in pest populations, grain temperature, moisture migration, spillage and grain residues.

It encourages general maintenance of the storage structure and surroundings in a sound condition thereby denying access to pests as far as practicable. It also removes potential sites for infestation developing in cracks and crevices within the storage fabric and the removal of disused or obsolete machinery, used and often infested bags and dunnage.Connell stated that, "Inspection and sampling work involves the exercises of judgement based principally on experience and an intimate knowledge of the nature, properties and physical behavior of grain.

Inspection, although considered of major importance, is more often than not underestimated or in fact neglected, because it is time consuming and precise methods cannot be adequately specified. General principles however are the same whether the object of the inspection for pests is for quarantine, grading or quality determination, or a scientific study; or whether it is performed in a grain warehouse, mill or processing plant, farm or ship. Information on the following is required:

- the presence of any pests
- if so, what species, population densities, location
- damage inflicted prior to the inspection
- potential for further development of these populations (their growth potential) or,
- whether the populations will eventually die out if conditions persist or they remain uncontrolled (seriousness)
- duration at safe storage of commodities or,
- the need for treatment or disposal. (urgency)

Subjective estimates must be critically avoided. While a numerical estimate of population numbers is seldom a practical approach, some quantitative estimate based on an arbitrary, internationally accepted and comparative scale must be pursued. It is important to realise that after adopting a thorough inspection programme, the inability to find any insect infestation does not automatically preclude their absence, and especially in the humid tropics of Southeaast Asia, the presence of low level populations now, will result in damaging population densities developing within four months. Continuous monitoring is therefore the essence of preventing serious deterioration of grain in storage.

Sampling and Inspection Programme

The objective of any sampling and inspection programme is to formulate the basis for future planning and action. Armed with precise and complete records of the history of stocks on hand, their present condition and presence of pests, the initiation of a feasible and economic control programme can be undertaken. Secondly, it forms an important base for determining its monetary value and whether the condition and quality of grain will satisfy the requirements of any potential buyer. Decisions based on a sample of grain that does not statistically reflect the overall condition of the bulk will result in erroneous conclusions being drawn concerning its fitness for human consumption. This problem is further compounded by the multiplicity of acceptance in quality standards; the nil tolerance for live insects in international export wheat as well as white flour in Western countries where it is condemned both in the eyes of the consumer and legally. Conversely, quite heavily infested grain or grain products maybe considered the norm in some developing countries depending on the economic status of the community.

Problems in Systems Assessment

It is considered a prerequisite that the inspector has a working knowledge of the insect species that are commonly occurring in the produce with which he is dealing. Recognition of the taxonomic and morphological differences. becomes increasingly important in establishing the status on endemic pests as well as identifying the introduction of any exotic species. Any species that is subject to quarantine regulations must be accurately identified by an appropriate authority. It is also advantageous for the inspector to have some knowledge on the biology, ecology and behaviour of various species which he may encounter, the differences between primary, secondary, field and associated pests, as well as beneficial parasites and predators.

Other practical difficulties have been elucidated by Howe, and can be summarised as:

- inaccessibility due to bad stacking or to inadequtely designed buildings and machinery
- sheer bulk of produce and size of premises
- variety of ecological niches and habitats
- insect behaviour: detection of latent or hidden infestation remaining inside the stack or bulk
- non-random or uneven distribution or insect populations

Problems associated with inspection techniques all focus on the desirability of investigating locales where insects are likely to abound, or to search for signs of the presence of insects rather than the insects themselves, particularly by recording rises in temperature with the use of thermocouples or thermistors inserted into bulk or built-in to bagged stacks.

Inspection and Sampling Techniques

Normal visual inspection of storage facilities and stored grain is subjective in nature, and therefore any results can only be recorded in a descriptive way. So long as the methodology remains uniform, direct comparison and therefore useful appraisals of different situations can be used. For more accurate and quantitative measurements it becomes necessary to sample the grain whereby the results can be interpreted by physical or chemical methods.

Oualitative Inspection Methods

Inspection is carried out with the objective of assessing the degree of insect infestation, although all forms of deterioration should be subsequently identified. Freeman first developed the need for standardisation of estimates of infestation into defined categories for all species and all products, and in some cases his estimate have been modified for the tropics where the degrees of infestation are somewhat higher than in Britain. Recording the results of a visual inspection have further been simplified by a useful shorthand notation as described by Ashman and has been widely adopted by inspectors (Table 1).

Table 1. Notation used for recording the results of a visual inspection or general inspect tion without taking samples and commodity inspection after drawing samples

1. Stack inspection	
C = Clear or none	No insects discovered in the course of a prolonged search.
F = Few or light	Small numbers of insects occuring infrequently or irregularly.
MN = Moderate numbers	Insects obvious, encountered regularly, sometimes forming small populations or aggregations.

Contd. . .

LN - Large numbers	Insects immediately obvious where large numbers are actively crawling over the entire surface of the commodity, i.e. stack or bulk.
VLN = Very large numbers	Insects extremely active and numerous that they are audibly present within the confines of the bulk or stack. Live insects or exuviae (cast skins) forming a continuous carpet around the perimeter of the stack or bulk.
2. *Storage inspector*	
C = Clear or none	No abvious insects or populations signs.
VF = Moderate numbers	Insects occuring regularly and and frequently, often forming populations but not obvious enough to be immediately noticeable.
LN = Large numbers	Insects immediately obvious on commencement of inspection crawling actively on walls and in other situations.
VLN = Very large numbers	Insects present in very large numbers, often forming dense populations on numerous surfaces as well as in any grain residues present on the floor, in mill augers, used sacks, dis-used machinery, bins, etc.
3. *Sampling inspector*	
C = Clear or none	No insects obvious on stacks or sacks or any of the samples. (Require protection from cross-infestation and regular inspection).
VL = Very light	Insects not obvious on sacks, or in sample of produce before sieving. < 20 insects per 90 kg sieved sample (Requires disinfestation in near future)
L = Light numbers	Between 20-300 insects per 90 kg sieve sample.
M = Moderate numbers	Between 50-300 insects per 90 kg sieved sample.
H = Heavy numbers	Between 300-1500 insects per 80 kg sieved sample.
VH = Very heavy numbers	> 1500 insects per 90 kg sieved sample

For example, living adults (A) Living larvae (L) Living pupae(P) Dead adults (a) Dead larvae (1) Dead pupae (p) General inspection may involve checking grain held at storage facilities for any obvious infestaion without drawing grain samples or it may involve looking for sources of residual infestation within the fabric of the storage structure or the immediate surrounds.

General Inspection of Grain

The following developed by the Ministry of Agriculture, Fisheries and Food Inspectorate, Britain, have international sanction and are therefore recommended for any general commodity inspection.

Inspection of Storage Facilities and Handling Equipement

The following categories apply to general inspections of the structural condition of the warehouse, silo or mill, or any handling and conveyance equipment that may offer suitable localities for endemic or residual insect infestations. In describing the species of insects present, it becomes necessary to employ the scientific generic and specific names, since the common names associated with insects sometimes vary between different countries. If the inspector is not absolutely positive of completely identifying species that are found, the uncertainty should be specified in the report, with every effort made to have the specimens in question identified by a specialist taxonomist.

The location where the species of insects were collected during the inspection should also be established in the report. They may not necessarily have complete distribution within the warehouse unless very heavy infestations are involved and this information will become an important consideration when embarking on any follow-up control procedures.

The main activities associated with a worth while inspection programme are making the effort for a prolonged search, observation and accurate recording of results. There exists no satisfactory alternative to be becoming actively involved either by getting into the grain, on to the stock, into the premises around handling equipment and physically searching for live insects, or the obvious signs for their presence. Because of the time-consuming nature of an inspection, the initial phase should concentrate on observing places where insects if present are likely to congregate. However, there are certain practices that maybe employed to make the job a little easier, most of which aim at stimulating the insects to come to you rather than the reverse.

Additional aids for visual inspection

During an inspection, a few bags should be opened at random and the folds of sacking and bag corners examined; Trogoderma granarium larvae and Tribolium castaneum adults are frequently found this way. Some bags should be at least lifted and set aside and the exposed surfaces of neighbouring bags quickly examined for adults and more carefully for larvae and pupae; such as for Ephestia cautella.

For the detection of light infestation, a more detailed examination is required. The following techniques may prove rewarding.

(i) *agitation of bags:* This is effective for low population densities of sitophilus granaries, S. orgzae and S. zeamais which will often walk out of sacks after they have been sufficiently disturbed. A long stick maybe drawn over surfaces of vertical stacks or they can be hit to activate small numbers of adult moths which are therefore more readily observed.

(ii) *the feel of grain in bulk:* Walking across the surface of bulk grain with bare feet may prove an excellent guide to its general condition. If it feels cool and free flowing chances are there is no cause for immediate concern. However, if a hotspot exists, this can be exemplified by solid caked patches indicative of high dust content and moisture migration with subsequent rises in temperature.

(iii) *traps:* Various traps have been designed to exploit the activities of many species of insects. Tube traps, consisting of a smooth inner surface and rough external surface can be inserted into bags to catch species such as Tribolium castaneum which are unable to escape by climbing the smooth inner surfaces. It is a useful cumulative trap, but is not effective in trapping Sitophilus sp. or Oryzaepilus surinamensis and may become ineffective where the webbing activity of moth larvae is apparent.

An extension of this simple design for trapping insects at varying depths in bulk grain has been developed in Canada.

Various forms of home-made, fly paper-type, sticky traps are commonly used to give an indication of the presence of flying insect pests at an early stage of infestation. Attractant traps, such as light taps (incandescent, flourescent or black light) as well as suction traps, can be helpful in large warehouses where suitable power is available to give reliable early indications of the presence of moths and beetles that fly readily. Traps that employ a sex attractant or pheromone as a lure offers a potential approach not only for estimating the degree of infestation but ultimately as control measure.

(iv) *artificial crevices:* Sections of corrugated cardboard can be placed between bags to attract pupating moth larvae. It can be examined after 24 hours for insects, but it is more applicable if the bulk is only lightly infested in which case examination may be done after 4 to 5 weeks.

Plank traps, consisting of two strips of wood 15 cm wide and 60 cm long, hinged together but held 3 to 4 mm apart, is use ful for Trogoderma granarium larvae and Tribolium castaneum adults. These are inserted in bags and left for several days before withdrawal and examination.

(v) *dead insects:* When residual protectants have been applied as a surface treatment and dead insects continue to accumulate, the conclusion might be reached that the treatment has been fully effective. Usually it indicates a source of live insects in the area, or from infested bags, some of which bemay obscured whithin the stack. If these insects are removed and the location marked for future reference, any further accumulation of dead insects indicates the need for further action.

(vi) *repellents:* Insecticidal formulations that possess a strong repellency action such as pyrethrin or synergised pyrethroids, can be particularly useful in exposing hidden insects in cracks and crevices. A light application will often stimulate insects to crawl onto exposed surfaces before they finally succumb to the insecticide.

(vii) *grain temperature and moisture content:* As mentioned earlier, it is offer more rewarding to investigate for the signs of insect presence rather than looking for live insects. Localised rises in grain temperature or moisture within a bagged stack or grain bulk are most important indicators of insect activity.

Dr. R. W. Howe has further discussed how the knowledge of insect pests and the physical behaviour and properties of stored grain should be used to help the inspector. He states, "Most storage insects are inconspicuous and secretive, and as a consequence are difficult to find. Nearly all storage insects are more easily found in dark premises because they are more active in the dark than in the light. They also lay eggs more readily in the dark.

The proportion of insects at or near the surface of produce varies with the insect species and the produce concerned. This is related to the size of the insect and its developmental instars and to the grain size of the product. Packing of stacks, diurnal rhythms, a tendency to stay near boundaries when brought to them by random movement, upward movement stimulated by disturbance, and outward movement stimulated by heating-all tend to bring insects near the surface of bagged stacks where the inspector has some chance of finding them.

Some of the reactions of insects to stimuli also help the inspector. Most prefer the dark; some are thigmotactic and collect in cracks between bags or under rubbish; most seek out wetter spots and many drink; yet others react to temperature gradients. Therefore, the inspector should examine dark places, the conical tufts of sprouting grain under leaks in the roof, the wet surfaces of bags and areas of produce known or thought to be wetter than the rest, and the tops of stacks especially those under metals roofs if Trogoderma is likely to be present."

Quantitative Sampling Inspections

The aim of drawing random samples of the commodity is

to determine the mean value and the variability of the level of infestation or contamination in any given situation. Ashman devised a tentative "sequel sampling" procedure, involving spear samples of the commodity taken at random and then examined by sieving. The procedure involves collecting a number of spear samples from several bags until a 1 kg sample is obtained, and examined for insects by sieving. Resampling occurs if low numbers are found and may involve a further three consecutive sampling occasions consisting of 3, 9 or 22 kilograms.

Degrees of infestation and need for appropriate action was as follows:

Spear sampling. There are many inherent problems with such a classification and could therefore be open to a variety of interpretations.

- Different insects preferentially attack different commodities and consequently inflict greater damage
- When insects are present in low numbers and are unevenly distributed, sampling spears are likely to give an inadequate assessment of the infestation, either grossly over or underestimated.
- The sampling size required to give an accurate assessment of the infestation using sampling spears is often laborious or time-consuming if the consignment is 10 bags then not less than 10 bags should be sampled at random.
- The sample taken is rarely a random sample, due to the difficulty of sampling from the central portion of the stack.
- Samples taken from individual bags is rarely a random sample or truly representative of the

condition within the bag. At least six more samples should be removed from each bag to make up the primary sample, a practice that is rarely adhered to.

– Sampling spears damage sacking and consequently create potential for more spillage.

Problems related to spear sampling are further exemplified in Figures 1 and 2 as depicted by Golob. Insects do not distribute themselves randomly or uniformly in any container of grain. They are most often found in pockets associated with dust, broken grain, and foreign materials towards the bottom or in areas of localised heating or wetting. It is very difficult to sample with a grain trieur or spear close to the peripheral margins of the bags, especially at the top and bottom. Therefore, a large population crawling on the bottom could be completely missed. Alternatively, small pockets of insects within the bulk maybe, by chance, included in a spear sample. For example, 6 sitophilus oryzae in a 500 g sample is not equivalent to 1200 individuals in a 100 kg. bag because of nonuniform distribution, and in fact the bag may contain less than 10 individuals altogether.

Typical spears for sampling bagged grains are represented in Figure 3. If used with due recognition of their limitations, they offer a cheap, simple and quick method in obtaining grain from bagged produce. To obtain more representative samples, a sectional probe should be utilised. These are available in sizes suitable for bags, or in larger sizes for probing deeper piles in bins, sheds, transport vehicles, etc. For larger samples in bulk grain, suction samplers consisting of a small hollow spear, hollow extension rods and suitable end attachments for the connection of an industrial type vacuum cleaner are adequate for drawing samples from up to 10 m below the surface. A more sophisticated version consisting of double construction interlocking aluminum extension tubes connected by flexible hose to an electrically driven vacuum pump is available. Samples are drawn into a collecting compartment, with the advantage that grain friction can be minimised by the vacuum and the probe inserted to any grain depth. The major disadvantage is its relative cost.

Other direct sampling methods involve either snaking of bags, coning and quartering and sieving.

(i) *bag snaking:* A number of bags maybe emptied by pulling the open bag backwards over the floor surface, allowing a small stream of grain to flow out gradually. Most visible insects will be concentrated in the latter portion and will be readily observed at the sides of the band.

(ii) *coning and quartering:* It is a simple and cheap method of obtaining highly representative samples but suffers from time and capacity constraints. The procedure involves tipping bags into the floor forming a cone, constantly mixing materials from the periphery to the apex of the cone, then spreading it evenly and dividing into 1/4, 1/8, 1/16, etc. subsamples depending on the volume required.

(iii) *sieving:* "Hand held sieves" are particularly useful in assessing the dust content and live insects from small samples. Different-sized mesh openings can be used for different particle size, or a combination of appropriate sizes can be used for mixed commodities varying in particle size.

"Sack sieves" have also been developed to sample an entire sack; the time taken can be between 5-15 minutes. The recovery of insects is dependent on insect species, time of sieving, slope of the oscillating sieve mesh and mesh size, but tests have shown better than 90% recovery of insects and is independent of population density. Insects pass through the vibrating screen along with fine materials which accompanies them.

For greater accuracy and representation, it is advantageous to take larger primary samples and then take subsamples to form a suitable working sample. Various methods can be employed, or by using specific apparatus designed for the purpose. These consist of the gravity mechanical types such as the Boerner conical divider, a simplified alternate channel box divider or the motorised centrifugal types such as the Gamet divider. TSPC developed the produce flow sampler (PFS) for taking samples from entire bags originally designed for sampling incoming loads from road transport. The time it takes is approximately 20 seconds for a 100 kg. bag, all flowable commodities can be sampled, and because samples are completely random, it is far more accurate than spear sampling.

Detecting Hidden Infestations

Most of the damage and weight loss caused by insects on grain are inflicted by the primary grain feeders. They are capable of penetrating sound whole kernels of grain and their life cycle is completed entirely within the kernel in which the egg is laid or entered by the first instar larva. The absence of any live adults of storage insects in grain samples separated by the previous methods, does not necessarily mean the absence of an infestation and consequently many methods have been devised to identify individual kernels that have become the home of the immature stages of the major insect pests.

Concentration of infested grains

Any detection method is greatly enhanced if the infested

material can be adequately separated from sound grain and hence reduce the number of grains and the amount of dust and broken kernels that have to be examined. It becomes desirable to concentrate the insect-damaged kernels as a preliminary step. For this purpose, flotation in an air stream or liquid maybe employed.

Flotation separation in liquids

The feeding activity of insect larvae progressively reduces the density of the grain, and by immersing the grain sample in a liquid of suitable specific gravity, the infested grains should float and the sound ones sink. At specific gravities between 1.050 and 1.190, the floating layer contains only infested kernels and aproximately 50 to 70% of all infested kernels were separated. Absolute separation is therefore unlikely, but the presence of hidden infestations can be estimated quite accurately; while a general indication of the severity of infestation will also be obtained.

Flotation separation in the air

A vertical column with a fan which produced a stream of air sufficient to float the grain sample was used by Milner. By progressively increasing the intensity of blowing, it was noted that virtually all insect-damaged kernels were removed in the first two fractions, from which no emergence had occurred. The detection of insect-damaged grain can then be a relatively quick and efficient operation, and may speed up the exit-hole inspection procedure in commercial samples by a factor of ten or better.

Projection separation in air

By projecting the grains of a sample through the air at an initially constant velocity, they should separate according to their relative density. Air drag and gravity then further separate the grain according to kernel size, shape and the surface texture which also determine the bulk density or test weight of any particular grain example. Infested kernels tend to fall short of sound kernels of the same size and shape. This separation has enormous practical implication. Wheat that has been designated as heavily infested can be separated into different bins reclaiming as much as 50% of the grain for food purposes rather than condemning the entire consignment for feed purposes.

Detection of Insect Infestation Methods

Physical methods

(i) *Acoustic detection:* This method provides an immediate answer if active immature forms are present, and has an advantage of not destroying the sample. A microphone, low-noise amplifier, and loud speaker or a cathode ray oscilloscope display tube are required, and to limit extraneous laboratory background interference, the sample and microphone needs to be insulated or alternatively, the grain can be directly linked to a piezo-electric crystal.

Vibrations are noted at several characteristic frequencies, e.g., 200 cycles sec^{-1} are associated with movement and dispersal, while frequencies round 1200-1500 cycles sec^{-1} are associated with feeding.

This method is potentially valuable for detection of insect activity within silos and other storage facilities, and possibly transport vehicles, provided extraneous noises can be successfully eliminated. The system can be modified for adequate sensitivity under field conditions.

Other limitations of acoustic detection are that quiescent stages (pupae) and eggs cannot be identified. However, a commercial unit has been developed by SASAD, Budapest, Hungary called the Insectofon (R).

Acoustic microscopy. involving the use of high frequency ultrasound which had different velocities in tissues of different densities could be developed for insect detection in grain.

(ii) *Grain Radiography:* This method is appropriate for detecting hidden insects at most stages of development and has been developed and used extensively under commercial applications in the United States.

The use of X-rays, discovered in 1895 by Roentgen, was generally restricted to the examination of high density materials. Claussen and Shehan developed a process for making beryllium as a window material on X-ray tubes, and enabled the use of radiography with low density materials such as grain.

Several researchers at Kansas State University pioneered radiography in its application to agriculture, and developed a method for detecting hidden infestation that was marketed by the General Electric company known as the "grain inspection unit". It suffered from being time-consuming and was not suitable for routine spection in of grain. Stermer developed a completely automated X-ray system to inspect grain on kernel-tokernel basis that was more reliable and practical for use. His study also resulted in the development of a procedure using fine-grain "mammography" film which obtained a contrast of about 75%, after only 2.5 minutes exposure. The film is examined by low-power binocular microscope (6x to 30x) with transmitted light, and obtained an efficiency of close to 100% with fully grown larvae and pupae and 80 to 90% with early instar and eggs.

The automated X-ray analyser requires four major phases.

- The automated X-ray analyser suffers from being an expensive piece of laboratory equipment, does not distinguish between live and dead insects and treating with K_2CO_2 as a constrasting agent was a lengthy process.
- Simplicity of operator has been enhanced in various films such as self processing (Polaroid (R)) and cassettes for the Hewlett-Packard Faxitron(R) X-ray unit. The problem of interpretation of the radiograph which makes the X-ray techniques so different has prompted the USDA Federal Grain Inspection Services to scrap it altogether as a quick inspection method.
- The low-intensity X-ray imaging scope (Lixiscope) developed by NASA, is a fully portable lightweight unit powered by a penlight battery, and may find application for grain inspection.

(iii) *Nuclear magnetic resonance*: The application is similar to the X-ray techique but suffers from the same limitations of time constraints, sampling efficiency and relative cost.

(iv) *Carbon dioxide evolution:* This method gives an accurate measurement of the total metabolic rate of the grain, and therefore cannot be specifically applied to insects. The method requires enclosing a quantity of grain in a gas tight bottle at 35°C for 24 hours, then drawing a sample of intergranular air and analysing it for percent CO_2 evolved. Dry uninfested grain is normally < 0.25%, between 0.3-0.5% suggests a light insect infestation (or a mc> 15%), and if the CO_2 evolution is > 0.5% in 24 hours, the grain is definitely unsuitable for storage without any further treatment.

A further development requires the detection of respired CO_2 by infrared absorption spectroscopy which is extremely sensitive, and is applicable to all stages except possibly the egg. The procedure requires a chamber containing the grain sample to be purged with a carrier gas, and then sealed during which time the CO_2 evolved by insect activity reacts a concentration sufficient for detection. This takes approximately one minute and is followed by flushing with a carrier gas for 2 minutes, and moves into the infrared detector by a pusle-flow movement. It is capable of picking up slight increases in CO_2 above the natural atmospheric background concentration of 300 ppm and can detect one 4th instar larva of Sitophilus oryzae (L) in 350 ppm grams of wheat.

Two commercial prototypes based on their original concept were developed by Technico Instruments Inc., USA, which can process three samples simultaneously, and by Horiba Instrument Inc. of Japan which is a smaller portable unit. The Lawrence Livermore Laboratories of the USA developed a miniaturised atmospheric carbon dioxide detector system (MACDS) which is extremely sensitive and portable. Other sophisticated detection systems that may find future application in the grains industry involve infrared radiation detection such as Far infrared (FIR) imaging, photoacoustic spectroscopy and thermal imaging with a pyroelectric vidicon.

All these detection systems are based on the principle that all physical bodies with a temperature greater than absolute zero have an infrared radiation spectrum which is a function of the body's absolute temperature and that living metabolising tissues generally have a higher temperature than its surrounding environment, thus giving off different radiation characteristics.

(v) Breeding out. Grain suspected of being infested maybe incubated thus allowing insects to complete their life cycle. The biggest disadvantage is the time factor since even under the optimum conditions of temperature and moisture content, at least 4-6 weeks will be required to breed out the full population of grain weevils and even longer incubation periods will be required for many other storage species.

(vi) Visual examination of exit holes is the simplest method. It has been standardised by the FDA of the United States that > 3 holes per 100 grains was cause for rejection. An experienced observer can also detect the presence of mature, late instar larvae and pupae of weevils in grain by changes in colour transmitted through the seedcoat. This only indicates the presence of an advanced infestation, but neither the degree nor the severity of the infestation.

(vii) Grain dissection will reveal any internal infestation, and gives a valuable indication of the stage of development of an infestation relevant for any impending control method. It is best done under a binocular microscope, and dissected with a sharp scalpel after the grains have been presoftened by soaking for 2 hours.

(viii) Cracking flotation has long been employed to determine the internal insect content of grains. Insect fragments are released after coarse grinding of the sample. Concentration of insect material is achieved. from an alcoholic solution by flotationseparation with the addition of mineral solvents. This enables microscopic examination and identification, and provides objective data on the number of insects present as well as an indication about the stages of development within the test sample.

The technique is complicated and laborious, requires the facilities of a grain test laboratory and sufficient skill and experience on the part of the technicians.

(ix) Electrical capacitance and resistance methods have been investigated by Wirtz and Shellenberger where changes in both these properties occur when insests are persent within the grain.

Additionally, temperature testing as mentioned earlier is of extreme value in detecting the presence of insects, as well as the effectiveness of applied control measures as indicated by a gradual decline in grain temperature compared to the ambient.

Chemical Methods

(i) *Alkali treatment:* This method is often referred to as the sodium hydroxide gelatinisation procedure which depends on rendering the seedcoat and endosperm translucent so that the more advanced stages of immature insects become visible. Tests by Keppel and Harris have shown that it has too many constraints to be adaptable as a quick and efficient method, although its reliability is unquestionable.

(ii) *Egg plug staining:* A soluble flourescent dye is used to stain the gelatinous plug secreted by female Sitophilus spp. to cover the egg cavity in the grain.

Grains are soaked in a dilute solution of 20 ppm followed by rinsing and examining the kernels under ultra-violet light for the greenish-yellow plugs. The degree of internal infestation can be then estimated by the number of egg plugs observed. This method is not particularly accurate, it is time-consuming, gives no indication on the stage of insect development and is only useful for weevil infestation.

(iii) *Uric acid test:* Measurement of the uric acid content of infested grain will give an indication of the past insect infestations which may have been concealed during processing. The method was found too insensitive to detect present infestations, and was only useful when population densities were high, in which case they were visisly obvious anyway.

(iv) *Chemical detection of insect phenols:* A test based on spectrophotometry analysis for the concentration of a hydroxyphenol occuring in insect cuticle, which produces phenolidophenol dyes when chemically treated with 2, 6- dichloroquinone chlorimide was proposed. Bailey stated that the method showed particular merit based on Food and Drug Administration evaluations, but at that stage required more work to perfect the method.

(v) *Ultra-violet light illumination:* Ashman stated that insects or insect fragments when present in finely ground commodities will appear red when stained in crystal violet and illuminated with ultra-violet light. Similar to the uric acid test, it provides useful information of any previous insect infestations in the grain before processing.

(vi) *Ninhydrin system for insect detection:* Dennis and Decker described a machine for detecting pre-adult stages of insects in wheat utilising the chemical ninhydrin. In practice, the grain and any internal insects are crushed between filter paper which has been impregnated with a 0.7% solution of ninhydrin in acetone. If the grains are infested, the amino acids contained in the body fluids of live insects are absorbed by the paper and chemically react with ninhydrin to produce a strong purple colouration. The machine developed was automated, with a sample flow rate of 300 kernels per minute, spot paper was previously heated to approximately 120°C. The test paper could be kept as a visual record and retained for more than two years. Unreacted treated paper loses some of its sensitivity with aging, resulting in a slight colour intensity reduction if the paper is used several months after treatment. This machine was limited because of its size, and a more versatile smaller prototype was developed at the Tropical Stored Products Centre.

Compared to the X-ray technique for insect detection in a series of laboratory trials, the AshmanSimon Insect Infestation Detector proved superior in many respects, with the following deserving mention:

1. Cost of the X-ray technique was substantially higher.
2. Both techniques provide permanent records.
3. Determination of infestation was simpler and quicker.
4. The detector will detect 5-14% of eggs and early instar larvae; 40-60% of moderate sized larvae, 80-90% of late instar larvae in small cereal

grains such as wheat, while the X-ray technique is incapable of detecting eggs and small larvae and is not significantly better in estimating mature larvae or pupae in grains.

5. Skilled technicians and expensive laboratory equipment are not required.
6. The machine is adapted to both manually cranked and/or electrical operation and is capable of dealing with a wide range of food grain of various particle size by altering the gap between the rollers.

Some problems that may interfere with the ninhydrin test have been enumerated by Bailey. Moisture content above 16% m.c. will give a faint general reaction, especially if contaminated by storage molds. Other forms of interference listed were 1) kernels damaged by cracking or checking; 2) kernels damaged externally by insect feeding or mechanical injury; 3) previously infested kernels where adult emergence has been completed; 4) kernels containing dead insects; 5) sooty or heat-damaged kernels; 6) kernels with a high free fatty acidity (FFA) value. These interfering reactions tend to produce distinct, sharply outlined spots.

When estimation hidden infestation of Sitophilus spp. in maize, individual kernels maybe infested by more than one larva, which produce an overlapping stain recorded as a single spot. The ninhydrin estimate in this case generally underestimated even though the grains have to be initially kibbled, but appears to be predictable and consequently a correction factor could be applied.

6

Ecology of Stored Products Insect Pests

Approximately 85% of all known animal species presently existing on earth are insects and it is generally assumed that only one fifth of all insects have been classified and named, meaning overall numbers of species would be approaching 4 million! Because insects have existed on earth ever since the Palaeozoic era (230-570 million years ago) and those resembling existing present day forms have been found in the rocks of the Permian period (230-280 million years ago), it is little wonder that due to their potential adaptation, diversification and immense numbers of different species, some have found an ecological niche in stored grain!

The advantages of small size and short generation intervals are important in their survival, when you consider that small animals need less food and oxygen to maintain them; small animals can find more room in a given area than larger ones and can find shelter where larger animals can find none.

Insects are members of the superclass Hexapoda, class Insects of the Phylum Arthropoda which also contains the arachnids, crustaceans, myriapods, pycnogonids merostomes, symphylans and trilobites. Insects have the following morphological characteristics as being representative of the class;

(i) invertebrate animals that have a cuticle hardened to from an external skeleton

(ii) body divided into obvious segments

(iii) three pairs of jointed segmented legs

(iv) most insects have bodies clearly divided into

- the head, which carries the mouthparts, the eyes, and a pair of antennae
- the thorax, which carries the 3 pairs of legs and 2 pairs of membranous wings (in most adults), and
- the abdomen, which contains the digestive and reproductive organs.

Like all animals, insects are classified into various groupings or taxa, the basic unit being the species defined as, "a group of inter breeding organisms which is reproductively isolated from other related groups." Systematic taxonomy attempts to classify animals according to either their evolutionary relationships or by morphological divergence (phenetic affinity). The following insect orders are most commonly encountered in stored products.

a) Coleoptera	Beetles-most of these are either primary or secondary pests, some are associated pests such as scavengers, fungus-feeders, structural pests or predators.
b) Lepidoptera	Moths-either primary or secondary pests, the larval stage inflicting losses.
c) Hymenoptera	(i) wasps-beneficial parasites or parasitoids on beetles and moths.
	(ii) ants-usually scavengers, but occasionally pests.

d) Psocoptera	Dustlice-scavengers and pests, particularly in Southeast Asia where they sometimes occur in prolific numbers.
e) Hemiptera	Bugs-some incidental pests but mostly beneficial predators on beetles and moths.
f) Diptera	True flies-some invaders or incidental intruders, some are scavengers and a few are predators.
g) Dermaptera	Earwigs-incidental intruders and scavengers.
h) Isoptera	Termites-structural pests in warehouses of wooden construction.
i) Thysanura	Cockroaches-scavengers which may occasionally approach pest status.
k) Orthoptera	Grasshoppers, crickets and locusts-most are incidental field pests, but the "house cricket", Acheta domesticus Linnaeus, occasionally becomes a nuisance indoors, eating almost any kind of food or refuse.
l) Siphonaptera	Fleas-bloodsucking ectoparasites of mammals and birds. Wingless, with complete metamorphosis. 1900 species of which Pulex irritans Linnaeus and Xenopsylla cheopis Rothschild are sometimes found in granaries and warehouses.

Nearly one thousand species of insects have been associated with stored products throughout the world, of which the majority belong to the Coleoptera (60%) and Lepidoptera (8-9%). Beetles and moths are advanced, holometabolous types of insects which have a life-cycle we describe as complete metamorphosis. This life-cycle has two very important features:

(i) Because an insect has an external skeleton that cannot grow or stretch, a growing insect has to shed its skin so that it can expand its size while the new skin is still soft. The development of an insect happens in several stages, and at the end of each stage the old skin is shed.

(ii) The immature insects (larvae) that develop from the eggs have evolved as rather simple stages designed for feeding and growing. They are completely different in appearance from the mature adults. Therefore at the end of the period of larval growth, the larva becomes a pupa. This pupa does not feed or move, but inside its skin the body becomes completely changed into the form of the adult. At the end of the pupal stage, the mature adult emerges from the pupal skin. The four basic stages-egg, larva, pupa and adult are completely different in appearance.

The life-cycle of beetles and moths is summarised in the following diagram. The larvae of all beetle and moth storage pests feed on the stored food. The adults lay eggs in large numbers during their lifetime. The adults of many important beetle pests also feed on the stored food, but the adults of some beetle pests and all moth pests are short-lived and do not feed.

Major Pests in the Cereal Production

Stored grain insects can be grouped according to their relative importance on a particular commodity in a particular geographic region. Species can be classified according to frequency and abundance on particular commodities, such that "major insect pests" are found frequently due to their almost cosmopolitan distribution and in large numbers. Examples are *Sitophilus oryzae Linnaeus* and *Tribolium castaneum Herbst*. "Minor insect pests" would then simply mean those species that are encountered infrequently and at relatively low population densities.

A more frequent and misunderstood classification is based on the insects capacity or ability to infest sound, undamaged kernels. On whole cereal grains, pulses and oilseeds we often distinguish between primary and secondary pests, which have the following distinguishing characteristics; primary pests are:

i. capable of successfully attacking, feeding and multiplying on previously undamaged grains;
ii. are adapted to feed on a narrow range of commodities;
iii. usually cause very distinctive damage;
iv. usual develop within the grains, and often complete their entire development within a single grain;
v. are selective in their egg-laying behaviour;
vi. often infest the ripening crop before harvest; and
vii. usually cannot develop on the same food if the grains are ground (milled).

Secondary pests are

i. not capable of attacking previously undamaged grains, but can only attack and breed in grains that have been damaged by primary pests, physical damage by bad handling, threshing, drying or intentional processing that removes or damages the seedcoat.
ii. usually attack a very wide range of commodities;
iii. usually cause non-distinctive damage;

iv. sometimes develop within grains, but never complete their development within a single grain;

v. do not usually have selective egg-laying behaviour;

vi. are very rarely found on the crop at harvest; and

vii. are usually capable of developing on the same food after it is ground.

It must be emphasized that primary pests does not necessarily mean more important, but simply the dynamic processes involved whereby secondary pests can cohabitate or follow primary pests and inflict serious and economic losses, especially under long-term storage. Controlling primary pests can reduce the potential damage caused by secondary pests. Some pests do not fall into either specific category, whereby some secondary pests can attack whole grain if the moisture content is high, some preferentially feed on the germ while others can breed on whole grains, but do so much more rapidly on damaged grains. A more convenient and meaningful classification of the relative importance of insect pests, can be best described in relation to both losses incurred on different commodities and the absolute need to implement control strategies to prevent economic loss.

Pests of major importance – Those that regularly cause significant losses of commodity or require applied control measures to prevent such losses;

Pests of Moderate Importance – those that occasionally cause significant losses or need control measures;

Pests of Minor Importance – those that rarely cause loss or need control but are often associated with a commodity.

Rare or Absent Pests – those that are rarely found in or are absent from a commodity.

Paddy

Unhusked rice (paddy) is usually considered more resistant to insect attack than most other cereals. Moisture content is particularly important and field infestations of *S. cerealella* are the main problem. This pest is of major importance in areas yielding over 80% of paddy production. Infestations occur particularly during after harvest but do not persist in storage. The initial attack, however, allows other species such as *O. surinamensis* and to a lesser extent *T. castaneum* to become established. These species may also be prevalent as a result of husk damage or breakage of grain during harvesting or threshing, for example from combine harvesting. Where infestations of *S. cerealella* do persist in storage, they occur primarily in the surface layers and particularly around the edges of bulk stored paddy. Bagged paddy is less liable to attack, especially if stored in close-woven bags.

S. oryzae and *R. dominica* may cause serious damage in parboiled paddy and in varieties less resistant to attack particularly at comparatively high moisture contents. *R. dominica* predominates in the drier batches when these are left undisturbed for long periods. There is considerable confusion as to the relative roles of *S. oryzae* and *S. zeamis*. *S. oryzae* appears to be the dominant species in the dry areas and *S. zeamais* in the warm moist areas. *S. zeamais*, for example, is not common in the Middle East or in India where *S. oryzea* is predominant, whereas in Southeast Asia, *S. zeamais* is particularly abundant in milled rice and maize. The occurrence of field infestation, the greater mobility or *S. zeamais*, the rapid turnover a stocks in mills and the associated production of maize appear a probable combination of factors that, together with high local temperatures and humidities, would favour this. However, it has been demonstrated in Indonesia that *S. oryzae* appears the dominant species on paddy.

Most other species are of little consequence except that occasional outbreaks of *C. cephalonica* and *T. granarirm* may occur.

Rice

Milled rice is particularly susceptible to insect attack. It is used to store the grain as paddy and process or mill it as required, keeping the storage period of the processed product to a minimum.

S. oryzae is the perdominant species. It has been recorded as a major pest in areas that yield 80% of rice production. *S. zeamais* has not received a high rating as a pest but, as with paddy, its real importance is probably much greater than appreciated because of its being confused with *S. oryzae* particularly in eastern Asia and in Central America and the Caribbean.

B. dominica has been recorded as a major pest in areas yielding approximately half of rice production but because of the limited storage periods usually involved, this species would not be expected to cause damage of the same order as *S. oryzae*. *S. oryzae*. *C. cephalonica* and *T. castaneum* have frequencies of occurrence similar to *R. dominica*, and would be of greater significance. *O. surinamensis* and *E. cautella* must also be taken into account as significant pests though they do not occur as frequently as the species listed earlier.

Rice bran is a commodity notoriously susceptible to storage pests. It provides a serious source of infestation in many rice mills- *T. castaneum* is the principal species involved.

Maize

World production of maize is of the same order as rice

and not substantially less than wheat. The export trade is second only to wheat but because maize has a greater association with subsistence farming and local animal feeding, a lower proportion enters international trade. The two grains together account for approximately 80% of the world export trade in cereals.

Maize, more than any other cereal, is subject to field infestation. The only other cereals in which field infestation is problem today are the sorghums and paddy where infestations primarily concern *S. cerealella* attacking unthreshed grain left drying in the field after harvest. Maize, in addition to being attacked by *S. cerealella,* is heavily attacked when standing in the field at early stage of ripening by *Sitophllus zeamais.* This species may complete one or even two life cycles before harvest. Such field infestations which are often accompanied by a complex of Nitidulidae, Lathridlidae, Mycetophagidae and sometimes in later stages species such as *T. castaneum,* reflect varietal characteristics of the maize plant such as husk length, and their effect may be augmented, sometimes independently, by damage from birds and field crop insects such as Heliothis. The problem may be accentrated by crib drying and storing of maize cobs.

As with sorghum, varietal differences in hardness of the grain also can have a marked influence on severity attack. *Sitophilus oryzae* is again the most important pest. *S. Zeamais* is considerably less important though its importance probably would have been somewhat not separating it from *S. oryzae* in some maize-producing countries where it is known to be a pest were taken into account. The pattern of occurrence of the two species conforms reasonably with what could be expected from climatic considerations-in the warmer often more humid areas, *S. aramais* is more important than *S. oryzae,* with *S.*

zeamais becoming less important as the humidity decreases, or the temperature decreases, Both species appear absent in very hot areas such as in the Sudan but in most instances the lower temperature limits for maize appear not to preclude either species being present at least as pests of minor importance. *S. granarius* can also be regarded as a significant pest of stored maize particularly in these cooler areas, and in storage of imported maize in countries which it cannot be grown.

S. cerealella is of comparable importance to the *Sitophilus* spp. As with sorghum stored on the head, maize stored on the cob is liable to reinfestation in storage increasing the potential of the species. The larger grain size and hence intergranular spaces of maize allow some movement of *S. cerealella* in shelled grain and thus increase the chance of reinfestation in grain bulks. This is not common with the smaller more tightly-packed grains but when it occurs infestations are usually confined to the surface and the periphery of bulks.

The occurrence of *Sitophilus oryzae* which is capable of attacking pulses has also been recorded by Pemberton and Rodriguez on carob pods from Portugal. Carobs are normally very susceptible to storage pests which predominanty attack dried fruits, since the pod contains a high sugar content. Certain strains of *S. oryzae* are also known to breed on other pules and legumes, such as yellow split peas.

Infestation

Cross infestation

This may be defined as the infestation of one commodity by movement of insects and mites from one commodity to another. This happens when sound grains are stored side by side with infested materials or by loading both products into the same hold of a ship or the same railroad car.

Residual infestation

Residual infestation results from attack by insects which have remained in the structure of the store, vessel or vehicle after the removal of a previously infested commodity. Grains that have collected in corners, cracks and crevices in walls, floors and ceilings of a warehouse, or in the holds of a bulk tanker or railroad vehicle support insect populations that are capable of reproducing in large numbers and infesting clean grain.

Residual infestation and the potential it creates maybe of greater significance than insects within the bulk, which can be effectively treated or processed before insect numbers develop into large populations. Once insects become established in stores or within the fabric of storage structures, they maybe extremely difficult or impossible to eradicate unless the entire building can be effectively fumigated. This would be relatively expensive and may not be practicable if a sufficient level of gastightness cannot be maintained within the storage structure. The cost-effectiveness of the treatment also may not be warranted if cross-infestation from adjacent storages (who are not prepared to carry out extensive pest control measures) cannot be prevented.

(i) Infestation arising from the use of infested containers: Bags and sacks and other repeatedly used containers are often infested by insects. Of these, jute bags are most often infested by flour beetles, warehouse moths, tropical warehouse moths or rice moths. The receival of infested stocks without inspection or the recycling used sacks that are not adequately cleaned and treated with either a fumigant or impregnated with a residual insecticide, further compound the problem and reduce the efficiency of control measures.

(ii) Intestation arising from mills

(iii) Combine harvesters, field bins, augers and transportation equipment: After use, this are usually left in the field for several months and pockets of infestation may occur inside the machinery.

(iv) Grain elevators and related conveyance machinery

Infestation by flight and crawling of insects

Some stored grain insect pests are capable of flight and infestations are frequently initiated in this manner. All moths and many beetles move to stored products by flight. Some species also crawl into new uninfested stored products, but migration by crawling is confined within narrow limits.

Field infestations have been dramatically reduced by the adoption of combine harvesting in many Western countries. In the Southern states of the US, cereals especially corn is at risk in the field to such pests as *Sitophilus zeamais, Sitotroga cerealella, Sathrobrota rileyi, Nemapogon granella, Araecerus fasciculatus* and Carpophilus dimidiatus among others, the first two being the most dominant and serious.

Field activity of *S. zeamais* and *S. cerealella* has also been reported in the Southern parts of the USSR as well as in the Northeastern states of USA following mild winter conditions. Both species commonly occur in maize grown in the coastal region of NSW, Australia, while *Sitophilus oryzea* infested 9-17% of sampled standing heads of wheat in Queensland during October and November.

Field infestations can be curtailed by growing corn varieties that possess tight husk characteristics, disposing of infested grain in farm storages before the corn reaches the silking stage in the field, early harvesting, application of recommended grain protectants, drying and shelling and storing corn in tight bins or containers adaptable for fumigation.

Both *S. zeamais* and *Rhyzopertha dominica* have been shown to infest sorghum in the field. Preharvest spraying with either Tetrachlorvinphos, Pirimiphosmethyl or Malathion during the dry or wet season harvest afforded protection is storage for 3 and 2 months, respectively, without any further post-harvest treatment.

Infestation may also arise from natural sources including nests of birds, rodents, spiders and insects.

Development Factors of Infestations

The likelihood of a pest infestation developing and the severity of the infestations are determined both by the location of the stored grain habitat and by the specific features of the habitat itself.

(a) Important characteristics of the stored grain 'habitat'

i) Size of the grain bulk is of critical importance. Small bulks are very much subject to the vagaries of their immediate surroundings. Other things being equal conditions tend to become more stable as the bulk grows larger-the effect of climate within a bulk is tempered and the relative importance of mass immigration and emigration declines due to the insulating properties of bulk grain. If it is not moved or aerated, conditions in bulk will be largely governed by the temperature and moisture content of the grain when it was placed in storage.

ii) Shelter offered—masses of grain kept in tightly sealed containers are less subject to the vagaries of climate and less susceptible to mass invasions by biological agents.

iii) Climate—local conditions of temperature and moisture have the profoundest influence on the susceptibility of grain to infestation and the course of events in these infestations. Generally pest control is much more difficult and grain deteriorates more rapidly in warmer-moisture areas than in drier cooler regions. But as pointed out in (i) and (ii) the effects of climate may be greatly altered both by the size of the bulk and the degree of shelter offered by the particular habitat.

iv) The duration of stay of the grain—if the grain only stays a very short time in the habitat, then organisms infesting the grain will be subject to considerable disturbance and the dynamics of infestation will be greatly altered. While ephemeral habitats may serve as sites for initiating infestation they are not places where pest numbers increase to any extent; rather these are often places which test the survival capacity of the pest. It is only in undisturbed sites, preferably sheltered and preferably with at least modest quantities of grain that infestations can develop to damaging levels.

v) Proximity of sources of infestation. Some habitats such as poorly-cleaned storages may have adjacent sites harbouring numerous pests which can quickly invade incoming grain. Other habitats may be far from any such reservoirs and hence, other things being equal, should present less difficulty from a control point of view.

(b) Important ecological characteristics of storage insects

i) In the protected environment of grain, insect mortality is low in the immature stage. As a result, their capacity for increase is great.

ii) As the temperature inside stores is usually higher and more stable than the external environment, the effective temperature is more conducive for the development of storage pests than for field pests. Furthermore, many of stored product pests have originated from tropical or sub-tropical regions, thus they are capable of reproducing continuously throughout the year.

iii) Stored grains are usually hard and are relatively dry. The important pests are adapted to feed on hard food and to survive on a very low water content in their diet. They are quite sensitive to grain moisture content and equilibrium relative humidity.

Both (ii) and (iii) are optimised in the humid tropics for the rapid development of stored grain pests, i.e., temperatures between 27 and 32°C and equilibrium relative humidities between 70 and 75%. As a result storage insects are able to develop at more-or-less the same rate throughout the year. Most of the important pests have generation times of between 3 and 6 weeks on most products in tropical stores, i.e. between 8 and 16 generations per year.

iv) Insects that are small in size are extremely suitable for living in stockpiles of grain. Light affects different species, and different stages of development, although they commonly prefer darkness. There are many of the adult species that have already lost their flying ability.

v) The movement of infested parcels of grain and grain products in international trade has provided an effective means of ensuring the widest possible geographic spread of storage pests since cereals were first harvested. The continued introduction of the major pest species into climates outside their normal distribution, ensures also that all these species may be found in all parts of the world.

Because of the bulk and relatively low value of cereals, international transport is almost exclusively by sea although rail and road transport may become significant in limited areas. The infestation which occur in ships originate either in the commodity before loading or from cross-infestation from residues of previous cargoes or other infested cargoes being carried. It is often difficult, however, to determine the precise origin of particular infestations.

Pests in Grain Community

Clearly the amount of damage done to a bulk of grain is dependent upon the number and kind of pests present in the grain and the time available for these organisms to do the damage. Accordingly it is not unreasonable to centre one's understanding of the events in grain around the population dynamics of the pest.

An insect population in bulk grain experiences a certain rate of population growth and the population acquires a particular structure. Different stages of insect may be more damaging to grain than others. Growth rate and composition are products of factors influencing survivorship of adult and developmental stages, rate of development to maturity and the egg laying schedules to females. The insect population is influenced both by its environment-physical and biological conditions in its surroundings-and by its own size and composition. Depending on the species, insects may feed on whole grains, frass from other insects, fungi, grain particles and dust or other insects. In addition, the metabolic heat of insects and the water released by them encourages their own growth and that of other pests. The chain of cause and effect is complete when insects influence the very factors which had affected their population growth.

Abiotic Environmental Factors on Pests of Stored Grain

(a) Temperature

Below a certain minimum threshold temperature insects do not complete development from egg to maturity and the pest population cannot increase. At temperatures only slightly above their threshold, say within 4-5°C, mortality rates are extremely high for virtually all stages of development. Most species do not multiply fast enough to become a pest until temperature is somewhere between 3-6°C above the minimum threshold for development.

The highest rate of oviposition usually occurs at a temperature about or just below the developmental optimum. Temperature has a profound impact on insect and mite locomotion, the net effect being that insects tend to move out of areas where temperature is unfavourable and relocate in more favourable zones. Temperature effects on a pest is not always direct. Different members of the ecosystem have different temperature optima, hence temperature influences processes such as competition, predation and ecological succession. Temperature is also the main variable determining the rate of population growth, insect activity, the liberation of heat and moisture and the production of frass and other waste products which lead to grain deterioration.

In general, the higher the temperature, the shorter the developmental period. The most favourable temperature for development and reproduction is about 28°C. Below 21°C and above 35°C reproduction almost ceases. Developmental period from egg to adult becomes longer and the number of eggs laid fewer, at lower temperatures. At higher temperature, the life cycle becomes shorter while the oviposition rate increases. Rice weevils, for instance, complete their life cycle in 25 days at 30°C while they take about 94 days at 18°C. Most insects grow more quickly at 29-32°C

At a temperature over 34°C, insects usually cannot develop. Lesser grain borers and Khapra beetles will however continue to develop at 34°C and 40°C, respectively. The lower temperature limit for insect development depends on the species. It is 14-15°C for the rice weevil while it is 18-19°C for the rust-red flour beetle. These limits are the most important factor affecting the geographical distribution of stored insect species.

(b) Moisture

Different developmental stages of insects, mites and fungi have different susceptibilities to aridity or excessive moisture. Generally the dormant stages-eggs and pupae for insects, eggs and resting stages of mites, and spores of fungi can best resist desiccation while active feeding stages may die out if conditions are too dry. Tolerance depends on the species-some insects may increase even at a R.H. of 10-30%, while miters and other insects require at least 65% R.H. and fungi at least 70% R.H.

Insects depend primarily on their food supply for moisture to carry on their life processes. Stored product insects can develop on food with moisture content as low as 2 to 14%. Moisture requirements vary with species. Generally, optimum grain moisture for development and reproduction is 14-18%. Most species do not develop below 10% m.c.; some exceptions are the Mediterranean flour moth and Khapra beetle can develop on food with negligible moisture content, weevils cannot live on grain containing moisture of less than 10% and lesser grain borers can subsist on grain with 9% moisture content while the confused flour beetles can live on 2% moisture grain.

Moisture and temperature interact to provide conditions favourable for multiplication and survival of stored product insects. As temperature is increased, insects are able to reproduce in grain of lower MC and when moisture is increased, they can multiply at lower temperatures.

(c) Gas concentrations

Metabolic activities of grain, but to a much greater extent of mites, insects and especially fungi may cause local reduction in the concentration of oxygen and increases in carbon dioxide. As grain settles, intergranular passageways become constricted or even blocked if dust or grain particles are present. Low levels of oxygen and high levels of carbon dioxide cause metabolic stress on insects, mites and aerobic microorganisms generally increasing mortality while lowering fecundity and rates of development. If oxygen should drop below 2.5% conditions become lethal for insects, but fungi can survive with levels lower than 1 % O_2.

Table 1. Lower limits of relative humidity, temperature, optimal range of temperatures and degree of hardiness of pests of stored grain, together with estimate of maximum rate of increase per lunar month

Insect	Min. Rel. Humidity %	Min. Temp. °C	Opt. Temp. °C	Maximum Rate of Increase +
Acarus siro	65	7	21-27	2,500
Ephestia kuehniella	1	10	24-27	50
Ephestia elutella	30	10	25	15
Ptinus tectus	50	10	23-25	4
Endrosis sarcitrella	80	10	24-26	30
H.pseudospretella	80	13	24-26	2
Sitophilus granarius	50	15	26-30	15
Sitotroga cerealella	30	16	26-30	50
Gnathocerus cornutus	40	16	24-30	15
Stegobium paniceum	60	14	25-28	7.5
Sitophilus oryzae	60	17	27-31	25
Ephestia cautella	25	17	28-32	50
Plodia interpunctella	40	18	28-32	30
Corycra cephalonica	30	18	28-32	10
Dermestes masculatus	30	20	30-25	30
Oryzaephillus mercator	10	20	31-34	20
Cryptolestes turcicus	50	21	30-33	50
Oryzaephilus surinamensis	10	21	31-34	50
Tribolium confusum	1	21	30-33	60
Lasioderma serricorne	30	22	32-35	20
Necrobia rufipes	50	22	30-34	25
Tribolium castaneum	1	22	32-35	70
Cryptolestes ferrugineus	10	23	32-35	60
Trogoderma granarium	1	24	33-37	12.5

Biotic Environmental Factors on Pests of Stored Grain

(a) Food

Grain itself is just one of several resources used by pests

for nourishment inaddition they may feed on damaged grains, grain particles and dust, fragments of insects and mites, including frass and cast skins, dead and living arthropods and fungi. Thus, fungus beetles and certain moths show marked preference for mouldy grains or frass from other insects; certain mites make their living by feeding on other mites, and flour beetles supplement their diet by feeding on the dormant stages of their own and other species.

Stored grains are an ideal food source for stored product insect pests, providing the essential elements required for continued growth and development. Levels of carbohydrates, proteins, fats and vitamins required varies with the species concerned. The Indian meal moth, warehouse moth, Mediterranean flour moth, and saw-toothed grain beetle cannot live in artificial diet containing carbohydrates of less than 20%, while the confused flour beetle and cigarette beetle can live on foodstuff with less than 2% carbohydrates.

The saw-toothed grain beetle can grow with 2-3% protein of artificial diet and hide beetles need at least 20% protein. Insects need 10 kinds of essential amino acids: arginine, isoleucine, leucine, Iysine, methionine, phenylalanine, threonine, trytophan and saline. Some Insects utilise fats in food materials as their source of energy, and some, such as the warehouse moths and tropical house moths, fats are indispensable. Their adults will partially or completely lack scales on their wings or fail to emerge from pupae if fats are not available.

Most of the vitamin B group such as thiamine (B.), riboflavin (B2), pyridoxine (B6), nicotinic acid, patothenic acid, biotin, folic acid, and choline are indispensable to all insects.

However, the saw-toothed grain beetle, drugstore beetle and cigarette beetle can live on food materials without some B group vitamins, not because they can dispense with these vitamins, but because they have in their bodies some intercellular symbiotic microorganisms which are able to synthesise and supply such vitamins. Vitamins A and C are not required by any of the stored products insects.

(b) Competition

At low and moderate population densities members of a given species will improve conditions for their fellows. Initial attacks on hard grains may make things easier for the next, while heat and moisture released by metabolic activity may push temperature and moisture into more favourable zones or at least encourage the growth of other species. Beyond optimal population densities, the scramble for food, for mates and for oviposition sites enhances mortality, lowers rates of development and leads to lower fecundity.

Members of various species residing in stored grain have considerable influence on the welfare of other consumer species. Thus as a larva *Tribolium castaneum*, the fulur beetle benefits from encounters with eggs of say *Ephestia cautella* the moth, which it feeds on, but as an adult the beetle may get injured by a larval moth. Insects and mites may lower the fecundity of the other species simply by mechanical disturbance-jostling-or by releasing noxious chemicals. Feeding and release heat, water, and waste by some consumers may pave the way for succession by other consumers, or species may engage in a scramble for resources causing such depletion that other consumers die out or are forced to leave the locality.

The competition for food is more severe between species which have the same nutritional and ecological requirements. Parasites and predators have, to some extent, helped in reducing insect populations. Some insects such as *Tribolium* spp. or flour beetles exhibit cannibalism.

Types of Damage

Direct damage

1. *Reduction or loss in weight:* The direct feeding of insect pest on stored grains results in food and weight losses. A rice weevil will eat 14 milligrams of a rice kernel weighing about 20 milligrams during its developmental period from egg to adult. However, this results in the commercial loss of the whole rice kernel.

   Weevils, which feed mainly on the carbohydrate portion of a maize, remove a considerable amount of the calorie potential but little of the protein and vitamins which are mainly in the germ and bran. In legumes, in which the protein and vitamins are more evenly distributed throughout the grain, infestation of beetles, which can be responsible for consuming up to 50% of peas and beans of which some 25% of the dry matter is crude protein, can cause a loss of about 12% of the available protein.

2. *Reduction or loss of seed viability:* A seed grain which has been attacked by a germ feeder will not germinate. Examples are *Triblim spp., Cryptolestes* spp., *Tenebroides niauritanieus, Ephestia* spp.

Indirect damage

1. Quality loss

a) *Nutrient loss:* When grains are attacked by insect species which feed selectively on the germ leaving the endosperm almost untouched, food loss is not apparent; weight loss is also small compared to loss of loss of vitamins, proteins, etc.

b) *Heating and spotlage:* Heating from insect infestation accelerates further infection by microflora and bacteria. This results in spoilage of grain which often causes more serious economic damage than loss in weight by insect consumption.

c) *Contamination, tainting or discoloration:* Contamination and tainting of foodstuffs with frass materials such as insect fragments, excrete, secretions, webbings and dusts also contributes to a deterioration in quality. Insect secretions or entomoxins, such as quinone, may also prove to be serious contaminants.

d) *Production of off-flavors and odors:* Cereal, after prolonged storage, but particularly when ground into a meal or flour or when infested by insects, show an increase in free-fatty acid content (FFA). If, prior to milling, the maize has been attacked by adults of Tribolium castaneum, the initial far of the resultant meal is much higher than that of uninfested meal, and the FFA rises to a much higher level.

2. Monetary loss

Financial losses are not simply in terms of that lost in reduction in weight, but also the downgrading or absolute rejection through the presence of live insects or signs of their activity. The cost of any applied control measures, as well as all the agricultural inputs invested in growing the grain, become important "hidden" costs if grain is consumed or destroyed by insects before reaching the consumer.

3. Loss of goodwill

A country loses its goodwill and earns a bad reputation if its produce are of low quality, insect infested and contaminated by toxins, insect, excrete, etc.

Strategy for Survival

Successful storage pest insects have adopted strategies that can be summarised as being opportunist. By a variety of biological mechanisms, all these species are capable of very rapid population growth when they invade a batch of grain in a suitable condition. Most pest species have potential rates of increase of between 20 and 60 times per month under optimum conditions (Table 1). We should consider what this means by thinking about a pest with a 40-times-permonth rate of increase: One gravid female invading the grain can lead to 40 individuals after one month, 1600 individuals after two months, 64,000 after three months, 2,560,000 after four months.

Fortunately, at some point in the development of the population, natural or artificial factors operate to control the increase. If pest control is not applied, the commonest natural limiting factor is competition for available food while other forms of competition or disturbance, presence of fungal growth and mycotoxins, parasites, predators and pathogens, pheromones and defensive secretions from other species, and cannibalism become contributory factors in suppressing insect populations.

Levels of temperature and moisture liberated by the development of an insect population often becomes excessive for the survival of the pests, eventually forcing the insects to relocate themselves elsewhere, thus increasing the hotspot to a point where adults can only survive in the superficial layers or on the surface. The very success of a species or species complex within a population may very well form the basis for its own extinction.

As well as being opportunists in taking advantage of abundant food placed in a store, most major pest species also have a survival strategy for the time when the food is removed in many cases, this involves the establishment of small residual populations which immediately have the potential to infest a new batch of grain entering the store. Another survival strategy found in many primary pests is their ability to disperse to ripening grain in the field. Finally, many species rely on human distribution and transportation to implant them in another storage locality. The use of second hand untreated sacks helps promulgate and distribute major storage pests, and is perhaps one of the major causes of cross-infestation.

Storage insects have become adapted to surviving in a low moisture environment relative to their field pest counterparts. Any significant changes in temperature and relative humidity (which are functions of the equilibrium moisture content of different grains) will have a critical effect on the rate and development of insects, as well as the composition or species complex constituting a pest population. It must be emphasized that the accepted level or "safe moisture content" to which grain should be dried does not prevent attack by insect pests, and in fact a relative humidity of 70% is well within the tolerance of most major pest species. Oftentimes in the humid tropics, grain possesses a moisture content somewhere within the 70-75% e.r.h. range. At the higher levels, the development of storage fungi is delayed, but has the deleterious effect that many major insect pests develop much more rapidly and many minor insect pests that are inhibited at 70% r.h. attain pest status.

If grain is not fully dried to the safe moisture levels, then

i) the slightly softer grain allows attack by many pests that initially cannot develop on sound, undamaged kernels.

ii) higher humidity becomes more favourable to a wider range of pests, and

iii) presence of slight but significant fungal growth allows rapid increase of certain pests that require a certain proportion of fungus in their diet.

Finally, it should be noted that drying must be uniform throughout a batch of grain. The presence of a few grains that are not fully dried can allow the rapid development of some insects consequently producing more moisture, and allowing the development of pests in grains that were originally dry but have absorbed the extra moisture. The infestation "hotspot" can thus expand from a very small volume and become a serious problem throughout the entire batch of grain unless some measure of control is implemented to reduce the population and disperse increased moisture.

Primary, Secondary and Associated Species of Stored Products Coleoptera

The beetles comprise the largest natural order in the Animal Kingdom with a total of no less than 250,000 described species, of which more than 600 have been associated with stored food products throughout the world. Some, through the agency and dispersal by man in international trade have attained cosmopolitan distribution and constitute the major cereal pests that attack stored cereals and grain legumes. Ever since man began storing food reserves in granaries during biblical times, various beetle species have become established pests and have continued till the present day as one of the most profound causes of grain loss in storage. Adults of the "cigarette beetle", "humpbacked spider beetle" and "weevils" were discovered in 1390 BC from the tombs of Tutankhamen, and no doubt the association of stored grain beetles with man extends even further back during the neolithic phase of human prehistory.

Characteristics of the Order

Small to large insects, forewings modified to form horny wingcovers or elytra, which meet along a straight mid-dorsal line. The hind wings are membranous, folded away beneath the elytra, some times reduced or even absent in Sitophilus granarius. The mouthparts are adapted for biting and chewing. The prothorax is larger and more developed than the other thoracic segments extending dorsally as a single shield called the pronotum. Metamorphosis is complete, i.e. egg to larva to pupa to adult. The larvae, have well developed heads and mouthparts and are of three main types (a) campodeiform; elongate and active with well developed antennae and mouthparts, often carnivores e.g. Carabidae, Staphylinidae (b) scarabaeiform; crescentic, bulkier and less active, usually herbivores e.g. Anobiidae, Bruchidae and (c) apodous scarabaeiform; grub-like, with reduced antennae and mouthparts which live within the food source e.g. Curculionidae.

The pupae of Coleoptera are mostly exarate but some are found within earthen cells and a few within cocoons (pharate) formed within the cast skin of the final larval instar. They are morphologically diverse and have invaded land, air and aquatic habitats with great success. Their hard cuticle is of particular value in surviving both biological and physical hardships as well as their ability to enter into facultative diapause under adverse conditions. Many are important to man as pests of growing crops, timber, skins and hides, textiles, furniture, structures and of stored foods and drugs. Some others, such as blister beetles (Meloidae) are directly injurious to mankind by their toxic secretions.

(F) cannot penetrate sound paddy husks in the early larval stages, but gains access through the presence of minute cracks. Species such as *R. dominica* and *Trogoderma granarium* Everts can be considered intermediate between the two classifications, since only the late instars can penetrate undamaged grains. It also appears that both *Sitophilus oryzae* (L.) and *Sitophilus zeamais* Motschulsky are also not true primary pests of paddy due to the female's inability to penetrate sound husks, and since feeding is directly related to oviposition, egg-laying is inhibited unless some physical defect exists. However, these defects are apparently not uncommon an observation based simply on the frequency of occurrence of weevil infestations in paddy, especially after further husk damage has been inflicted by the feeding activity of the lesser grain borer or *Sitotroga cerealella* Oliver, the Angoumois grain moth.

Very little attention has been focused on some secondary pests such as *Lophocateres pusillus* and *Cryptolestes* spp (especially *C. pusillus* (Schon.) and *C. ferrugineus* (Stephens)), which are commonly found in moderate to large numbers on paddy in the humid tropics, sometimes even outnumbering the acknowledged primary pests. They are often assumed to cause little damage but it would seem apparent that such large and abundant populations maybe utilising whole paddy, since the early instar larvae of these beetles would be certainly capable of gaining entry through defects necessary for the entrance of *R. dominica*.

Once husk and kernel damage has been initiated by the primary grain feeders, the extent of damage caused by these and other secondary pests of milled rice would be quite significant, especially in long-term storage. The physical condition of a grain mass, its temperature and moisture content considerably affects its suitability as a habitat for a given species because each has its own particular temperature and humidity limits and optima. The activities of pests such as *S. oryzae* and *R. dominica* often change the temperature and moisture content of bulk grain causing heating and moisture migration and thereby initiating a succession of consumers within the storage ecosystem resulting in grain deterioration.

The major coleopteran families that we are concerned with in stored products entomology are listed in Table 2 (in alphabetical order).

Major Stored Products in Beetle Families

1. Anoblldae

Adults are small, subcylindrical (wood boring) oval or nearly globular with the prothorax more or less covering the deflexed head. The stored products members of the family have 11-segmented antennae with a loose 3-segmented serrate club. The elytra completely cover the abdomen and there are 5 visible abdominal sternites. All tarsi are 5segmented with segments 1-4 decreasing in length.

A family which are damaging to wood i.e., *Anobium puncatatum*, the furniture beetle, and *Xestobium rufillosum*, the death watch beetle. The family is related to the Bostrychidae and the Ptinidae and although cosmopolitan, is best represented in temperate regions. The larvae are scarabaeiform, fleshy and have enlarged terminal segments; and can be disinguished from allied members of Ptinidae by possessing minute spinules on some of the folds of the thorax and abdomen.

Table 2. Major coleopteran families.

Family	*Common Name*	*Number of described species*		
		Total (World)	*Stored Product*	
Anoblidae	Furniture, tobacco and biscuit beetles	1000	15	(2)
Anthicidae	Incidental pests	1800	2	
Anthribidae	Coffee bean borers	2400	1	
Bostrychidae	Wood borers	434		(3)
Bruchidae	Bean weevils	1200		(11)
Carabidae	Ground beetles (predatory)	25000	(4)	
Cleridae	Chequered beetles (mainly predatory)	3400	10	(4)
Colydiidae	Incidental pests: biology unknown	1400	3	
Cryptophagidae	Fungus beetles	800	25	
Cucujidae	Flat grain beetles	500	15	(4)
Curculionidae	"Weevils"	60000	30	(4)
Dermestidae	Hide and Leather beetles	600	55	(13)
Histeridae	Carrion feeders	2500		(1)
Languriidae	Mexican grain beetle	400		(i)
Lathridiiae	Plaster beetles	520	35	(7)
Mycetophagidae	Fungus beetles	200	5	(2)
Nitidulidae	Sap feeding beetles	2000	16	(4)
Ptinidae	Spider beetles	500	24	(8)
Silvanidae	Secondary grain feeders	400	15	(4)
Staphylinidae	Rove or cocktail beetles; Biology variable	27000		
Tenebrionidae	Darkling beetles	15000	100	(14)
Trogossitidae	Secondary grain feeders (germ)	600		(2)
(Ostomatidae)	Some predatory			

Two species *Lasioderma serricorne* and *Stegobium paniceum* are of importance as stored product pests.

(a) *Lasioderma serricorne (F.):* The tobacco or cigarette beetle. Small (2-2.5 mm), oval, brown or dark brown beetle. Antennae serrate throughout; absence of line of punctures or striae on the elytra and lack of spinules on abdominal larval segments distinguish if from other anoblids infesting stored products.

The eggs are laid singly in crevices of folds in the substrate. The fleshy larvae, which are distinctly hairy, pass through four instars. Newly hatched larvae cannot attack undamaged grain but enter small holes in search of food thus gaining entry to packaged goods such as cigarettes. The larvae pupate, when mature, within a rather flimsy cell amongst the substrate of within a grain or bean. The adult, lives only about 25 days producing some 100 eggs at 30°C and 70% R.H.

When humidity is not limiting the life-cycle can be completed in the range of 20°C to 37.5°C although the optimum is about 32.5°C. High humidities retard development and increase mortality as do humidities lower than about 70% R.H. By day, the adult beetle shuns light, hiding in crevices, but flies readily at dusk in warm conditions and is attracted to artificial light.

L. serricorne attacks an extremely wide range of stored products throughout the tropics and subtropics and is particularly damaging to tobacco, cigars, cigarettes, cocoa-beans, spices and various seeds such as coriander and caraway, cereals and cereal products.

(b) *Stegobium paniceum (L.). The drugstore beetle:* Small (2-3.5 mm), cylindrical, light brown beetle which resembles *L. Serricome* but whose last three anthennal segments are very long and elytra striated.

The life-cycle is similar to *L. serricorne*: the egg is followed by 4-6 larval instars, the last of which constructs a cocoon within which it pupates. After eclosion, the adult remains in the cocoon for several days before emerging and does not feed. At 20-25°C and 60-80% R.H., each female lays about 75 eggs and has a life-span of about 40-90 days.

The life-cycle can be completed over 15-34°C, development taking 40 days at about 30°C and 60-90% R.H. At 20°C development takes about 100 days at 90% R.H. and 130 days at 60% R.H. The adult cannot fly and dispersal therefore depends on passive distribution during the movement of goods. *S. paniceum* is cold-hardy and can survive winter conditions in temperate regions.

S. paniceum is a cosmopolitan pest of more temperate than tropical climates infesting almost any dry animal or plant products, particularly in pharmacies where it eats drugs and in premises where it attacks breakfast foods, biscuits, etc.

2. *Anthribidae*

Adults infesting or associated with stored products resemble Bruchid bean weevils. The antennae are 11-segmented, thread like or filiform with a loose 3-segmented club, rarely with a compact club. All tarsi are 5-segmented of the third segment which is nearly always broadly dilated. Mostly of tropical distribution, associated generally with old wood dead branches and fungi.

The eggs are laid within the endosperm. The larva is spodous scarabaeiform similar to Sitophilus. In maize of highs moisture content, the female lays about 80 eggs in a life-span of some 60 days. The optimum range for development is 28-32°C but the speed of development and mortality are much influenced by the moisture content of the grain. Thus, at 27°C and 80% R.H. the life-cycle is completed in 43 days in maize and 66 days in cocoa. The lower limits for development are about 20°C and 60% R.H.

A. fasciculatus, which is now more or less cosmopolitan, especially in the tropics where it attacks a wide range of stored commodities, such as coffee and cocoa beans, nutmegs, ginger, maize and dried fruit, particularly when such produce is not properly dried. It will also attack the fallen fruits of the custard apple, bananas and cotton balls.

It is a strong flier and is frequently seen in cornfields in Southern USA on exposed and damaged ears. It lays eggs in the soft kernels and continues to breed after the corn has been harvested and placed in storage. However, it is considered a minor pest since it does not cause significant damage once the corn becomes too hard to be attractive. In the humid tropics, it is considered a major pest of cassava, attacking whole tubers and chips while they are still being dried.

3. Bostrychidae

Prothorax coarsely tuberculate and concealing the deflexed head. Antennae terminate in a loose 3 segmented club.

A cosmoplitan family of mainly wood-boring beetles which, particularly in the tropics and subtropics, attack felled timber or dry wood and, occasionally, unhealthy trees. Besides *Rhizopertha dominica,* only *Prostephanus truncatus* (Horn), the larger grain-borer, which attacks corn, grain and tubers in the Southern states of USA, central and south America and *Dinoderus minutus* (F.), the bamboo borer, which attacks bamboo, grain, flour, tobacco and spices in the tropics and warm temperate areas are of any significance as pests of stored products.

P. truncatus previously regarded as having a restricted distribution to the Americas, has now been recorded in Africa, in the Tabora region of Tanzania and has now spread to Neighbouring Guinea, Togo, Kenya, Burundi, Benin and Ghana. The "greater grain borer" or "Scania beetle", is capable of boring through the husk sheath of maize, and losses reported in 3-6 months storage were 10% and sometimes as high as 30% in sample that had a moisture content of only 11% m.c.

Rhyzopertha dominica (F.). The lesser grainborer. Small (2.5-3 mm) dark brown cylindrical beetle with head concealed by thorax which has many small tubercles. The elytra have clear rows of regular punctures. The female lays her eggs in clusters on the grain in cracks and crevices or singly amongst the grain dust and frass produced by the adults. The newly hatched, white, cylindrical larva, may enter the grain directly or feed amongst the dust until it reaches the third of the four larval instar when it usually enters grain and eventually pupates. More than one larvae may be found within a single grain. Adults live for 2-3 months and females produce 200 to 400 eggs depending on conditions. Under optimum conditions of population density, temperature, moisture and presence of damaged and cracked grains, the adults can survive almost 5 months and produce approximately 600 eggs. The adults feeds voraciously, leaving characteristic holes with ragged edges in the grain, and produces floury dust and frass.

The life-cycle can be completed over 20-38°C, although the optimum is between 32°C and 35°C. Development is completed in 25 days at 34°C and 84 days at 22°C (both at 70°C R.H.) and ceases below 18°C. Moisture levels have a considerable influence on the biology of the pest, where few eggs are laid in grain of less than 8% moisture content although, in contrast to grain weevils, oviposition is little inhibited until moisture content is less than 9%. The adults are strong fliers and move freely from one store to another. Within a grain mass the species disperses rather slowly but is generally found in the driest part of the grain mass.

R. dominica is particularly well adapted to breeding in warm dry climates and is of particular importance in countries such as Australia and India. It is most important in wheat, rice, paddy, millet and of lesser importance in sorghum and is known from other cereals such as maize. Of little importance in milled products.

Bruchidae

Adults are small,- strongly convex beetles, densely clothed in hairs which are usually coloured and form distinctive patterns. The last abdominal segment is large, more or less vertically inclined, and conspicuous beyond the shortened elytra. The eyes are large, and with exception of the genus Caryedon, the eyes are u-shaped or emarginate. The antennae are 11-segmented without a club, usually filiform with some segments compressed while in the males of some Callosobruchus, they are pectinate or flabellate. The tarsi are distinctly 4-segmented with the third segment broadly dilated and deeply emarginate. The hind femora are somewhat thickened or strongly enlarged and each bears one or more teeth. Bruchid larvae on hatching have three thoracic legs but after the first moult they become apodous. Despite their common name, the bean weevils do not possess the elongate rostrum or snout of the true weevils.

Several species of bruchids are primary pests of stored beans, peas and other legumes.

Multivoltine species

(a) *Acanthoscelides obtectus (Say), Dried bean weevil:* Adults are form 3.2 mm to 4.0 mm long and grey in colour with light and dark markings on the wing covers. The hind femur has one large and two small teeth which are of taxonomic importance in this genus. This beetle is an important primary pest of pulses, especially of Phaseolus spp. beans. In all but the cool temperate regions of the world it attacks beans ripening in the field before harvest as well as in store. The larva develops entirely within a single grain, eventually forming a pupation chamber just under the seed coat, at which stage the infestation becomes evident as a characteristic circular 'window' on the surface of the grain. The adult which emerges from the chamber is short-lived and does not feed, and the female lays its eggs rapidly; the eggs are laid freely among the grains, in contrast to other pest bruchids which cement their eggs to the seedcoat. A generation takes about 4 weeks at 30°C and development is possible in temperatures down to 15°C.

(b) *Callosobruchus chinensis (L.), "Cowpea beetle":* Antennae of the males are pectinate to strongly pectinate, whlist on the female they are subserrate. The apical tooth on the inner carina of the hind femur is long, narrow, parallel sided and blunt at the apex. Adults generally 2.5-3.5 mm in length with typical white markings on the scutellum. This bruchid beetle is an important primary pest of pulses in tropical and subtropical climates, originating from the orient but now found in many other areas. Pulses are often infested in the field before harvest as well as in storage.

The larva develops entirely within single grain, eventually forming a pupation chamber just under the seedcoat, at which stage the infestation becomes evident as a circular 'window' on the surface of the grain. The adult which emerges from the chamber is short-lived and does not feed, and the female lays its eggs rapidly; the eggs are cemented firmly to the outside of the grain and the larva hatches through the 'floor' of the egg straight into the grain. After the larva has hatched the eggcase is clearly visible as a small creamish-white spot on the surface of the infested grain. The other major pest species of Callosobruchus have a similar life-cycle and cause nearly identical damage. One generation is completed in 25 days at 30°C, and females lay more than 100 eggs, usually in less than one week.

(c) *Callosobruchus analis (F.): C. analis* differs from all other members of this genus by having a tooth on the inner carina of the hind femur which is relatively small and sometimes absent. The eyes tend to be flattened and less prominent. Adults range in size from 2.5-3.5 mm. This bruchid is a primary pest of pulses in many parts of tropical Asia and has been reported on grain, soyabeans and groundnuts from India, Burma, Hongkong, Indonesia but rarely E. Africa. Generally less common than *C. chinensis,* but causes identical damage. Optimum conditions for development are 30-32°C and 70-90% R.H.

(d) *Callosobruchus maculatus (F.), "Southern cowpea beetle"*: The antennae of both sexes in this species are subserrate. The elytra are brown with two medio-lateral black spots; apical tooth on inner carina of hind femur straight. Slightly larger varying from 3-4.5 mm. Adults live for 710 days only, the female lays about 80 eggs, and one generation takes 21 days (32.5°C and 90% R.H.). This bruchid beetle is an important primary pest of pulses in tropical and subtropical climates, probably originating from Africa but now nearly cosmopolitan in all warm regions of the world except India where it is not common. Pulses are often infested in the field before harvest as well as in store. Certain types of pulses, particularly cowpeas, are more susceptible to attack by this pest than are other pulses. The life-cycle is similar to that of *C. chinensis,* and the damage caused is identical.

(e) *Callossobruchus rhodesianses (Pic), "Cowpea beetle".* Adults 2.5-3.5 mm in length; antennae of males serrate while on females sub-serrate. Colour of apical half of elytra black or nearly black in females, brown in males. Apical tooth on inner carina acute and curving towards apex. This bruchid is a primary pest of pulses in central and southern Africa and has not been recorded outside this continent. Pulses are often infested in the field before harvest as well as in pulses have dried. The life-cycle is similar to that of *C. chinensis* but is completed in 26-28 days at 30°C and 70% R.H.

Zabrotes subfasciatus (Boheman)

This bruchid beetle is an important primary pest of several pulses but is particularly associated with the many varieties of common beans (Phaseolus vu/garis). It is particularly common in central and South America but is also found in several other tropical or sub-tropical regions. The pulses are often infested in the field before harvest as well as in store. *Smallish bruchids* (2-2.5 mm long) with long, filiform antennae, black with basal segments reddish yellow. The dorsal surface of adults usually variegated dark and brown while the base of the pronotum has a large patch of whitish hairs also exist as a broad transverse patch on the elytra.

Caryedon serratus (= C. gonagra F.). "Groundnut borer".

The eyes of this species are large and not emarginate. The hind tibiae are strongly curved. Adults are larger (4-7 mm) and cosmopolitan in distribution. This beetle is a primary pest of groundnuts in many parts of the tropics. It can be a serious pest, especially on groundnuts that are inshell. Infestations normalty start in the groundnuts soon after harvest, often while they are being dried. The larvae spend most of their life within the pod but when they are fullygrown they leave the seeds and cut exit holes in the pods. Pupation occurs within a thin paper-like cocoon that is usually formed on the outside of the pod but occasionally protrudes through the exit hole. The adults generally live and lay eggs within the surface layers of a bulk of groundnuts: up to 25 cm into the bulk in groundnuts in shell, much less in shelled groundnuts.

Cucujidae

Small adults (1.5 to 2.5 mm) which are flattened, reddish-brown, II-segmented filiform antennae which may display a thickening of apical segments in some species. The prothorax possesses an entire, sub-lateral ridge on both sides which are more or less parallel. Torsal segmentation are 5-segmented in females while in males the front and mid-tars are 5 segmented and the hind tarsi are 4-segmented. Some members are often predatory but in Cryptolestes, the genus of importance in stored products, predation and grain feeding both occur. They closely resemble each other in appearance, while determination of species is based on characteristics of the genitalia and antenna: body length ratios of the males. (*C. turcicus* are as long or longer than the body, *C. pusillus* are only 2/3 as long as the body while *C. ferrugineus* are only 1/2 as long as the body).

Cryptolestes ferrugineus (Stephens): The flat or rusty grain beetle. Very small (1.5 mm) reddish-brown flattened beetle with long filiform antennae. The eggs are placed in crevices within the grain or dropped loosely. The elongate larva has pronounced tail-horns and feeds preferentially on the germ of the grain. It passes through four instars and pupates in a gelatinous cocoon which is usually covered in food particles. Cannibalism occurs under crowded conditions. After a pre-oviposition period of two or three days, the female lays approximately 400 eggs and lives for 6-9 months at 32°C and 70% R.H.

The life-cycle can be completed over the range of 20°C to 42.5°C and the optimum for both development and population increase is about 35°C, when development is completed in 21 days. Development is retarded and mortality increased at low humidities but the rate of population increase is still considerable at 40% R.H. This species is cold-tolerant and can overwinter without difficulty in very cold conditions.

The adult flies actively at temperatures above 21°C and warm grain, i.e. above 30°C is preferred. By virtue of its wide tolerance of temperature and humidity, *C. ferrugineus* is a cosmoplitan and important secondary pest of cereals, cereal products and oilseeds in temperate, sub-tropical and tropical regions and is often associated with or follows infestations of primary pests.

C. pusillus (Schonherr), The flat grain beetle: The optimum conditions for increase are about 35°C and 90%

R.H., development being completed in about 27 days under such conditions. It is not tolerant of low temperatures or of low humidities less than 60% R.H. A cosmopolitan pest attacking a similar range of commodities to *C. ferrugineus* but only abundant in the tropics. It is apparently unable to survive in sound uninjured grain, but is commonly associated with the more vigorous primary grain pests such as the rice weevil. It is also a scavenger, the larvae feeding on the remains of dead insects, and is most often associated with grain and meat that is in poor condition.

The larvae are particularly attracted to the wheat germ, and in infested grain, many kernels are found uninjured except for the removal of the germ.

C. turcicus (Grouvelie): Optimum conditions for increase are about 28°C and R.H., when development takes about 38 days, but the life-cycle can be completed between 17°C to 37°C at 90% R.H. At lower humidities survival is reduced and approaches zero at 50% R.H. The species is cold tolerant and is most abundant in flour mills in temperate regions where it is often confused with *C. ferrugineus* which occupies the same niche in the sub-tropics and tropics.

C. pusilloides: Optimum conditions for increase are about 30°C and 90% R.H., when development is completed is completed in about 23 days. Development is completed over 15°-35°C at 90% R.H. However, survival is reduced and approaches zero at 50% R.H. and is restricted to tropical and sub-tropical regions in the southern hemisphere where it attacks a similar range of products to *C. ferrugineus*.

Curculionidae: The Curculionidae or true weevils form the largest single family in the Animal Kingdom. Many species are pests of growing crops of many kinds throughout the world, and approximately 30 species have been recorded in stored food products, three of which are of extreme importance. Adults may be distinguished from all other stored products species by having the head produced in front of the eyes to form a well defined snout, the antennae are geniculate and clubbed, and all tarsi are 4-segmented. The larvae are apodous scarabaeiform, stout and slightly curved, and are creamy white with a pale brown or yellowish head.

Three species of Sitophilus are important pests of whole cereals, however, they are of little significance as pests of milled cereals because the larvae require a hard substrate in which to develop. As primary pests, these weevils are of additional importance in that their entry into whole grains provides access for secondary pests such as *Tribolium* spp. The role of the immature stages in the formation of 'hotspots' and in moisturemigration is also noted.

Sitophilus granarius (L.), The granary weevil: Small (2-4 mm), polished, uniformly chestnut brown to black weevil. Thoracic punctures oval and rather shallow. Male distinguished by shorter and more deeply pitted rostrum than female. The females bores a hole in a grain with her mandibles, lays her egg at the bottom of the hole which is then sealed with a gelatinous plug.

After four moults, the larva pupates within a single grain. If several eggs are laid within a single grain usually only one larva reaches the pupal stage due to cannibalism of the supernumerary eggs. Newly emerged adults spend some days within the grain before chewing their way out. Providing that temperature and moisture levels are favourable, the female lays about 250 (maximum 360) eggs during her life-span of about six months.

The life-cycle can be completed over the temperature range of 12-32°C, the optimum lying between 26°C and

30°C. Development is completed in 26 days at 30°C and 144 days at 15°C (both at 70% R.H.). The moisture content of the grain greatly influences both oviposition and the development of the immature stages. Thus, eggs are seldom laid in grain of less 9% moisture content and are most numerous in grain of 14-16% moisture content.

The larvae produce much metabolic heat and moisture during their development where the heat generated may contribute to the formation of 'hotsports' with spot temperatures of 38-42°C forcing the adults to migrate to cooler areas.

The adults have no functional wings under the elytra and their dispersal is therefore passive. *S. granaries* is typically a pest in rather cool areas because the developmental threshold and temperature optima are lower than those of other *Sitophilus* spp. A wide range of cereals are attacked particularly wheat, oats, barley and rye.

Sitophilus oryzae (L.), The rice weevil: Small (2-3 mm), brown to black weevil with two paler red dishbrown patches on each elytra. Prothoracic punctures round and rather deep, except for a smooth narrow strip extending down the middle of the upper side. Punctures on the elytra are narrowly separate, as compared to *S. granaries*, which are widely separate.

Life-history and biology essentially as for *S. granaries*. However, the female is more fecund and lays about 380 eggs during her lifespan of about five months. The life-cycle can be completed over the temperature range of 14-34°C; the optimum for development and survival is 27-31 °C i.e. somewhat higher than the granary weevil. Development takes 25 days at 29°C and 220 days at 15°C (both at 70% R.H.). Few eggs are laid in grain of less than 10.5% moisture content and mortality of the immature stages are increased in grain with less than about 13% moisture content.

The heat and moisture produced by the immature stages is of considerable importance in the development of 'hot-spots' and moisture migration. The adults have functional wings and their dispersal depends on flight as well as passive dispersal in trade. In contrast to *S. granarius*, the rice weevil with its greater tolerance of temperature exceeding 30°C, is typically a pest of warm temperate and tropical zones. It attacks wheat, barley, paddy, rice, millet, sorghum and maize amongst other grains but is of little importance in milled products.

Sitophilus zeamais Motschulsky, The maize weevil. Small (3 mm) but somewhat larger than *S. oryzae* from which it cannot only be distinguished by examining the genitalia. Often confused with *S. oryzae* or referred to as *S. oryzae* 'large-strain'. Although generally darker than the rice weevil the variation that exists within each species makes separation on external characters alone rather inaccurate.

Life-history and biology essentially similar to *S. oryzae* but is about 1.5 times more fecund than *S. oryzae* with a life-span of five months. *S. zeamais* will oviposit in much moister grain, e.g. ripening maize, than *S. oryzae* and ovipostition is inhibited in grain of less 12.5% moisture content.

The life-cycle is completed in 37 days at 25°C and 110 days at 18°C (at 70% R.H.) which is somewhat longer than in *S. oryzae*.

S. zeamais is an active flier resulting in many field infestations in areas adjacent to infested storages. *S. zeamais* occurs throughout the warm humid areas of the world and will attack a wide range of cereals although it is particularly a pest of maize. It will breed in rice and wheat but a true assessment of its pest status is confounded by the confusion that exists in early reports as regards to the separation of the species.

Caulophilus latinasus (Say), The broadnosed weevil (= *C. oryzae (Gyllenhal))*: Small (3 mm) dark-brown broad-nosed weevil. The life-cycle is essentially similar to the members of *Sitophilus* spp. The female lays an average 136 eggs over four months-the total life-span being about five months. Development is completed in 26 days at 28°C and 76 days at 17°C.

7

Pest Management for Greenhouse Crops

Insects and diseases are a major challenge to greenhouse production. IPM is an important tool in the management of these pests. The primary goal of IPM is to optimise pest control in an economically and ecologically sound way. IPM involves the integration of cultural, physical, biological, and chemical practices to grow crops with minimal use of pesticides. Monitoring, sampling, and record keeping are used to determine when control options are needed to keep pests below an economically damaging threshold. Pest management, not eradication, is the goal of IPM.

IPM is a simple, practical, and, most important, flexible way to manage insects, mites, diseases, weeds and vertebrates. Integrated pest management is adaptable to all greenhouse-grown crops and involves specific techniques to manage pests. These techniques are:

- Monitoring or scouting programme
    a) individual plant inspection
    b) yellow, blue, and hot pink sticky cards
    c) indicator plants

- Pest identification and life stages
- Record keeping to identify trends and direction for your pest management programme
- Exclusion techniques to prevent pests from entering the production area
  a) Insect screens to exclude aphids, whiteflies, and thrips from entering through doors and ventilating systems
- Cultural practices to prevent problems
  a) soil testing
  b) sanitation
- Biological controls, living organisms used to reduce the incidence of pest organisms
- Insect growth regulators, insecticides that interfere with normal insect development or the molting process
- Chemical controls
  a) proper choice of pesticides
  b) proper timing of pesticide application
  c) proper application procedure

It is important to understand the life cycle and behavior of insect pests in order to develop an effective control strategy. Knowledge about the weak link in a pest's life cycle can help growers choose the most appropriate control strategy. There are several practices that will increase the success of an IPM programme:

- Cover all soil floor surfaces with concrete, black plastic or weed barrier.
- Use resistant varieties of plants.
- Keep people and "pet plants" out of crop areas as much as possible.

- Pasteurise growing medium.
- Keep doors closed.

Greenhouse Insect and Mite Control

Major insect and mite pests of greenhouse crops include aphids, thrips, whiteflies, fungus gnats, mealybugs, shore flies, spider mites, leafminers and scales. In addition to damaging the crop, some insects vector diseases like tomato spotted wilt virus.

Crop Scouting and Trapping

To detect early infestations, a crop scouting programme that includes both sticky trap cards (usually yellow) and visual inspection is critical. Scouting should be done once a week, and more often after an infestation is detected. Regular scouting is also necessary to monitor the efficacy of control measures. A hand lens is a useful tool to detect live pests as well as signs of pest activity - e.g., frass (feces), cast skins, honeydew, etc.

Monitoring records can be kept on paper or in a computer. (Computers are usually better for producing graphs, which show trends more easily.) State and federal regulations will soon require that all greenhouses that apply pesticides keep records on what was applied. If plant injury symptoms appear, the grower can see quickly what chemicals or biocontrols have been used, and how the environment affected the crop . And, thorough records are invaluable in negotiations with suppliers for compensation for problems traced to receipt of diseased stock.

At a minimum, records should include:

- Minimum and maximum temperatures for each day
- Counts from sticky cards, changed weekly

- Counts of pests on the plants, including stage of growth (egg, immature, adult)
- Plant growth and development based on measurements of selected, flagged plants
- Root health based on weekly check of random plants
- Growing medium pH and soluble salts of random plants
- Specific crop observation (height, leaf color, bud development, etc.)

Sticky Ribbons

Sticky ribbons are long sticky insect strips that are hung throughout the greenhouse as a means of reducing insect populations of whiteflies, leafminers, aphids, fungus gnats, and thrips. Unlike sticky traps that are used for monitoring, the primary purpose of sticky ribbons is to reduce the numbers of flying insect pests by simply catching them.

Sanitation

Sanitation is key for controlling pests in greenhouses. The goal of sanitation is to eliminate all possible sources of the pest. Weeds inside and near the outside of the greenhouse can harbor pests. It's best to pull the weeds inside the greenhouse rather than spray them, since insects may survive the spray and migrate onto crops. Bag all weeds and dispose of them outside the greenhouse.

In addition, a 10-30 foot vegetation-free zone around the outside perimeter of the greenhouse-especially near vents and opening-can provide a dramatic decrease in pests. A heavy-duty geotextile weed barrier covered with bark mulch or gravel can provide a pleasant vegetation-free zone, and eliminate the need for herbicides. Plant debris from previous crops can also be a source of both immature and adult pests. Clean up all debris from previous crops and dispose of infested plants, or any infested growth. Ideally, the greenhouse should be thoroughly cleaned and left empty for one week prior to beginning the next crop. This enables removal of all pest stages, and starves any remaining adults. Closing up the greenhouse when it is empty in summer will increase the temperature and help eradicate pests.

Inside the greenhouse, a clean stock programme should be in place. This includes temporary quarantine and inspection of all plants upon arrival from other greenhouses, and regular monitoring of stock plants used for propagation. If a separate section of the greenhouse can't be dedicated to this purpose, flag all incoming plants. All new plant material should be thoroughly inspected (with a 10X hand lens) for the presence of pests to ensure that no infested plants are introduced into the greenhouse. Workers in the greenhouse should avoid wearing yellow clothing, since many pests are attracted to this color and may hitch a ride on the fabric from one greenhouse to the next.

Screening

Insect screens physically exclude the entry of lightweight, airborne insects like aphids, whiteflies, and thrips from the greenhouse through doors, cooling pads, and ventilation units. Although the tiny pores of insect screens prevent entry of insects, they can also impede the flow of air. Some specialists say the area the screen covers should be three times that of the area covered (doors, ventilation) to facilitate sufficient air intake. To accomplish this, screen houses designed to enclose both the cooling pads and greenhouse entranceway can be installed.

Biological Control

Biological control is the use of living organisms to control crop pests. Biological control of greenhouse insect pests can be achieved through release of biocontrol agents like predatory mites, pirate bugs, soil-dwelling mites, and parasitic insects.

Implementing a biological control programme in a greenhouse is management intensive and requires more knowledge on the part of the grower than do traditional pest control programmes. Proper species identification is very important before a control programme using predators or parasites is initiated. Release rate, timing, placement, temperature, and pesticide use also influence the success or failure of biological control efforts. Rigorous monitoring is necessary for proper timing of biocontrol agent releases, as is a reliable supply of the natural enemies.

Knowing the lifespan of the beneficials selected is important too, since fewer releases are required if sufficient numbers of parasites or predators are maintained. Some species of beneficials live only a few days and therefore must be released biweekly. Other species live for several weeks; consequently, they are released less frequently.

The level of control provided by a natural enemy will also vary with crop species. It is therefore more difficult for bedding and other container-plant growers to implement biocontrol, since they produce a wide variety of crops. One of the most important things to keep in mind when using biologicals is that even residual insecticide can harm them. This happened to a grower who tried to use *Encarsia formosa*, a parasitic wasp, on his poinsettias. The previous crop had been sprayed with a pesticide two months before applying the wasps, but all the wasps were killed by the residual.

Biologicals also allow more thorough coverage than spraying and a more efficient use of labor, since workers don't have to wait for re-entry into the greenhouse. Generally, the best time to release beneficials is at the first sign of a pest infestation. Early morning and dusk are the optimum release times. Suppliers of biocontrol agents can provide technical assistance and help growers get started in a biocontrol programme. When requesting assistance or ordering beneficials from a supplier, the grower should be able to provide information such as: what the pest is, how much square footage is affected, crop rotations, fertility programmes, and pesticides that have been used in the greenhouse.

The Canadian greenhouse vegetable industry-which is about four times the size of the U.S. industry-is largely based on biological controls. In 1995, tomato growers in British Columbia spent 50 cents per square meter per season on biological control, primarily on the parasitic wasp *Encarsia formosa*. There are several items to keep in mind when using biological controls:

- No single pest control method is 100% effective.
- This method often involves more work at first than chemical control, and it may require changes in production methods.
- Biologicals are often highly susceptible to pesticides. When choosing pesticides, select those with the shortest residual life and the highest specificity.

- Since many biologicals work slowly, they are best used when pest numbers are fairly low.
- Most predators and parasites perform best at moderate temperatures (65-85°F) and humidities (60-90%).
- If the greenhouse is allowed a dormant period (either very hot or very cold), the beneficials will die.
- If the pest level is very high when beneficials are first introduced, they probably will not provide enough control.
- The effectiveness of the same beneficial may be different on different plants. For instance, beneficials are often less effective on hairy-leaved plants like tomatoes.
- If all the plants leave the greenhouse, so will the beneficials.
- If all the pests are destroyed, the beneficials will starve.
- Some plants exude substances toxic to beneficials.

Insect Growth Regulators

Insect growth regulators (IGRs) are another least-toxic pesticide control option for pests. IGRs typically kill insects by disrupting their development. They have a complex mode of action that precludes insects from rapidly developing resistance. IGRs can work in one of several ways: 1) they can mimic juvenile hormones, so that insects never enter the reproductive stage of development; 2) they can interfere with the production of chitin, which makes up the shell of most insects; or 3) they can interfere with the molting process.

IGRs usually work through ingestion, so good spray coverage is essential. They generally don't affect non-target species-such as humans, birds, fish or other vertebrates. For most IGRs there are minimal re-entry restrictions. IGRs typically take several days to have an effect on pest populations. Because IGRs do not affect mature insects, adult beneficials released into the greenhouse after an IGR application are not likely to be affected. Use of IGRs is generally prohibited by organic certification organisations because the products are synthetic.

IGRs can sometimes be used in conjunction with biological control efforts and may provide growers with a "safety net" should beneficials fail to keep the pests below economically damaging levels.

Biorational Pesticides

The integration of biorational pesticides (also known as least-toxic or biopesticides) in greenhouses is often necessary in addition to cultural and biological control measures. When the use of a pesticide is necessary, materials should be selected that are least harmful to the predators and parasites released into the greenhouse. Insecticidal soap, horticultural oils, and the bacterium Bacillus thuringiensis are examples of insecticides that can be safely integrated into a biological control programme. The advantages of biopesticides over conventional chemicals are their selectivity to a targeted pest, lower toxicity to beneficial insects and greenhouse workers, and shorter re-entry intervals (REI).

Enhanced control with pesticides is achieved with thorough spray coverage. Wider plant spacing and removal of dead lower leaves improves pesticide coverage and pest control.

Beauveria Bassiana

BotaniGard™ (from Mycotech) and Naturalis-O™ (from SePro) are the two commercial formulations of *B. bassiana* that are available for greenhouse use. *B. bassiana* is a naturally occurring fungus that attacks a wide range of pests-including aphids, whiteflies, thrips and spider mites. *B. bassiana* has been found to be compatible with predators such as *Encarsia* spp., *Eretmocerus* spp. and *Chrysoperla* spp.

Spinosad

Early in 1999, Conserve SC™ from Dow AgroSciences was registered for use on greenhouse ornamentals. The active ingredient comes from the soil-inhabiting *actinomycete Saccharopolyspora spinosa*. Conserve is effective against many different species of caterpillars, leafminers, and thrips. Experiments performed at the Ohio State University showed that conserve provided excellent control of Western flower thrips.

Hot Pepper Wax and Garlic Extract

These products are actually insect repellents. Hot Pepper Wax™ contains paraffin (refined wax) and capsaicin (the "hot" in hot peppers). Another study at Ohio State showed that Hot Pepper Wax did not control Western flower thrips, but was effective against two-spotted spider mite. Products containing garlic also work to repel insects.

Greenhouse Disease Control

The greenhouse climate is ideal for the development of plant diseases. An integration of cultural practices, environmental control, biological control, and natural control products will be needed to prevent widespread outbreak. Many fungicides are also toxic to beneficial organisms, and should be avoided if possible. Alternative disease control techniques include the use of disease resistant varieties, disease-free seeds and plants, well-drained soil, air circulation, weed eradication, humidity control, sanitation, disease-suppressive composts, compost watery extracts, and microbial antagonists.

Disease control may be classified into two approaches: 1) those aimed at the root environment, and 2) those aimed at the aerial environment.

Root Environment

Soil disinfection (i.e., sterilisation) is an important part of soil-borne disease control when raising vegetables by the ground culture method or when soil-based potting mixes are used. Soil-borne diseases include damping-off (Pythium and Rhizoctonia), black root rot (Thielaviopsis), and several other root rots and wilts caused by Fusarium and Phytophthera. Potting mixes based on compost, peat moss, vermiculite, perlite, and bark are typically pathogen-free and do not require prior sterilisation.

Sterilisation involves heating the soil to 212°F for 30 minutes, a process that kills most organisms in the soil. Chemical changes also occur in the soil and may affect plant nutrient uptake to the following crop. Pasteurisation involves heating the soil to only 160°F for 30 minutes. While most harmful pathogens are killed at this temperature, many beneficial soil organisms survive. Thus, pasteurisation is the preferred method in organic programmes.

Chemical biocides, electrical heat, steam heat, and soil solarisation are the primary methods of soil disinfection in greenhouse production. A fifth method of disease suppression is biological control. Soil fumigants such as methyl bromide are, of course, restricted in organic production.

Electrical heat treatment, which is done inside a steel chamber surrounded by heating coils, is limited to treating about a cubic yard of soil at a time. Its primary use is in the pasteurisation of small batches of sand and soil for potting mixes.

Accordingly, steam pasteurisation and soil solarisation are the two most viable options for sterilising greenhouse soils or large volumes of soil-based mixes. Biological control is complementary to these two methods.

Steaming

Steam was the primary method of soil sterilisation in the greenhouse industry prior to the emergence of soil fumigants. Steam heat is highly effective and environmentally safe. Equipment and fuel costs are expensive, however, and treatment between crops is labor and time consuming. These are the chief reasons many growers shifted to soilless mixes.

There are three methods of steaming in common use today :

- Tarping an area and piping in steam for 6-8 hours, heating and sterilising the top 8 inches of soil;
- Pumping steam into subsurface drainage pipes, sterilising the top 2 feet of soil; and,
- Negative pressure steaming, where pumps pull steam applied at the surface through pipes buried 2 feet deep and 10 feet apart.

Solarisation

Soil solarisation is the process of tarping moist soils with clear polyethylene to trap solar radiation and raise soil temperatures to levels lethal to most pathogens and weed seeds. Solarisation is most effective when applied for at least 30 days in midsummer. Two layers of polyethylene, separated by fillers (i.e., pvc pipes or 2'x 4's) spaced every few feet to create an air space, increases the efficiency of solarisation.

Solarisation in greenhouses is a proven means of soil sterilisation, and is practiced worldwide in Japan, Israel, Greece, France, Italy, Belgium, Portugal, and Spain, in addition to the United States and Canada. Solarisation can also be used to pasteurise sand or soil intended for potting media, or to treat used media. A simple technique developed in Florida is to fill a black plastic trash bag with media, which is then sealed in a transparent plastic bag. The double-wrapped media is placed in the open on asphalt or concrete and spread to a uniform depth of 3 inches.

Biological control of pathogens in greenhouse soils and potting mixes is accomplished through natural control via cultural practices, and secondly, through applied biocontrols. The introduction of biocontrol agents, or antagonists, is the next form of biological soil disinfection. Biological fungicides are a promising alternative to synthetic fungicides.

There are several new biocontrols available for suppressing root diseases. One product is RootShield™, a biofungicide that controls root diseases caused by *Pythium, Rhizoctonia, Fusarium,* and *Sclerotinia*. RootShield contains the fungus Trichoderma harzianum and is sold either as granules or a drench. SoilGard 12G™ is another biofungicide that controls *Pythium, Fusarium, Rhizoctonia, Phytophthora* and *Thielaviopsis* (when co-applied with sulfur). The active ingredient is the fungus *Gliocladium virens*. SoilGard is sold as granules that can be incorporated into the soil mix. Another product is MycoStop™, sold as a wettable powder. MycoStop contains the actinomycete Streptomyces griseoviridis and controls *Fusarium*, although it may control other root diseases.

Cultural practices that promote soil health include crop rotation and use of tilled-in crop residues and green manures or organic amendments brought in from off site. The beneficial effect of crop rotation on plant health and yield is probably due to changes in soil microflora. These shifts result in increasing numbers of beneficial microorganisms (including nitrifying and other useful bacteria, antagonists of pathogens, etc.), and in decreasing numbers of noxious ones (pathogens and antagonists of beneficial microorganisms).

In some greenhouses, a short-term break crop-such as annual ryegrass, oats, buckwheat, or rapeseed-may fit into the rotation and be used to increase organic matter. Another option is to rely on organic amendments or green leaf manures brought in from off site. Amendments such as fresh residues, composts, and manures result in soil improvement through a proliferation of microflora.

Green leaf manures are field-grown cover crops that are chopped and harvested green, then transported inside the greenhouse to be incorporated into the soil. Many studies have shown that crucifer residues from rapeseed and oilseed radish cover crops produce toxic gases that suppress soilborne pathogens. Additionally, soil solarisation, following incorporation of cruciferous residues, increases the efficiency of soilborne pathogen control.

The role of composts as a slow-release nutrient source in organic production is well established. Recently, the role of composts as amendments for the control of soilborne plant pathogens has increased interest in similar use in conventional agriculture. In container production, disease-suppressive composts are commercially available in both peat and soil-based media. Natural disease-suppressive potting mixes-based on composts-are available through several commercial sources. These products may, however, contain starter fertilisers and wetting agents which would restrict their use in organic certification programmes. Contact the manufacturers for clarification on these ingredients. Some companies will blend a special batch that excludes restricted ingredients, therefore making them suitable for organic production.

8

Safe Grain Storage

Storage capability allows continuous supply of materials for processing or for distribution. On the other hand, it is undesirable because it means that huge capital investments in storage facilities will be required. Moreover on reception in storage, the product is subject to pest infestation and natural product deterioration. Most agricultural commodities are produced seasonally. Their harvesting is normally done during a short period of two to three months while their consumption is constant throughout the year. For this reason, storage becomes necessary.

The primary aims of storing food commodities are normally as follows:

1. To effect a uniform supply of food throughout the the year, either for domestic use or for export.
2. To provide a reserve for contingencies such as droughts, floods, and war.
3. To speculate on high prices either in domestic or in the export market.

Storage in the Rice Post-Harvest System

Storage is a critical component of the rice postharvest

system. In graphic form, the relationship of storage with other operations in the systems is as shown in Figure 1. As shown in the diagram, there are two points at which the storage operation is taking place - after procurement and just before distribution. Procurement is dictated by production in terms of timing quantity and quality. It is generally predictable and therefore the paddy storage operation can be programmed.

Local distribution is more or less fixed in terms of the constant rate of consumption. Export on the other hand is to some extent unpredictable. Delays in ship arrival and variation of quality specification by buyers demand that allowance in milled rice storage capacity has to be provided. Between the two storage operations is the milling operation. These three operations make a problematic trio to coordinate and control. Their functional relationship is illustrated in a model shown in Figure 2.

Elements of the Grain Storage System

The primary concern in storage is the safety of the product. For perishable crops, the product quality and quantity have to be maintained or deterioration has to be minimised. To achieve this purpose the five aspects of storage has to be considered, attended to and understood, namely:

1. The stored product
2. The storage structure
3. The environmental factors
4. The storage pests
5. The personnel involved

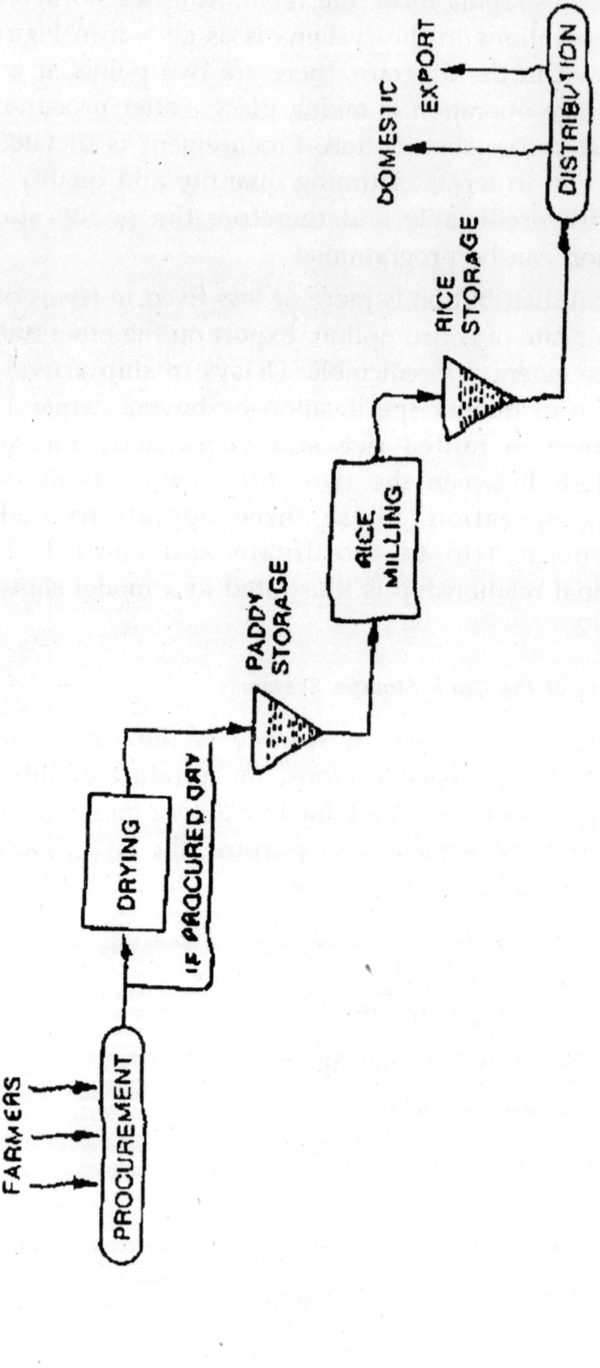
FARMERS
PROCUREMENT
DRYING
IF PROCURED DRY
PADDY STORAGE
RICE MILLING
RICE STORAGE
DISTRIBUTION
DOMESTIC
EXPORT

Figure 1

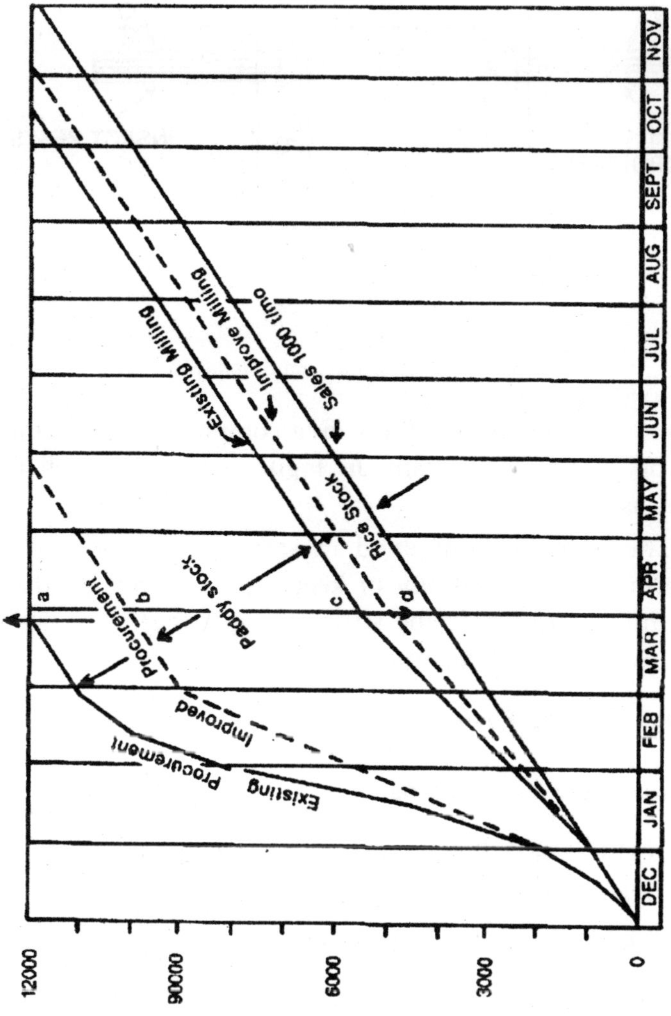

Figure 2. Model of Paddy and Rice Storage Capacity Needs with Different Potterns of Milling and Procurement

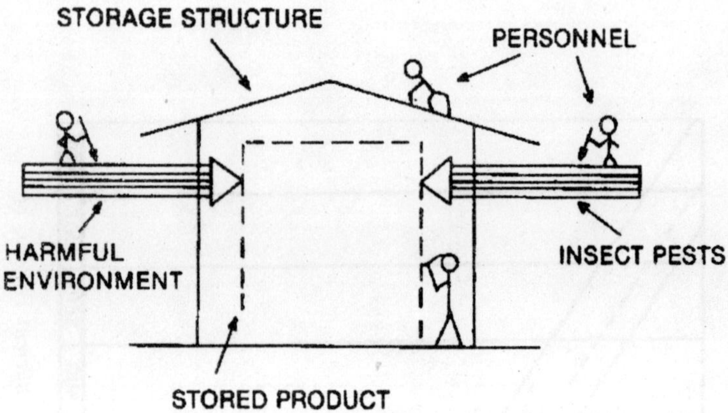

Figure 3. Components of the Grain Storage System

The interrelationship of these five components or factors may be shown graphically in Figure 3 or may well be stated as follows:

"The personnel involved in storage operation have the primary responsibility of keeping the stored product safe and secure inside the structure against storage pests and environmental factors.

Storage Lossing Factors

Mechanical and Environmental

1. Handling damage
2. Handling spillage
3. Moisture stresses and movements

Infestation

1. Bacteria, molds, and fungi
2. Insects

3. Rodents
4. Birds

Biochemical Processes

1. Vitamin loss
2. Fat acidity
3. Natural respiration

$$C_6H_{12}O_6 + 6O_2 - 6CO_2 + 6H_2 + 677 \text{ Kcal.}$$

Prevention of Storage Losses

Losses in storage can be minimised or prevented by adopting any or combination of the following:

Chemical Control

1. Insect control
2. Mold, fungi and bacterial control

Biological Control

1. Predators and parasites
2. Entomogenous fungi
3. Entomopathogenic diseases
4. Varietal resistance

Physical Control

1. Air conditioning (temperature & R.H. control)
2. Drying (moisture control)
3. Controlled atmosphere (gas concentration control)
4. Aeration
5. Heat disinfestation
6. Irradiation

Proper Design of Structure

1. Weather tight
2. Rodent and bird proof
3. Gastight

Sanitation

1. Cleaning of storage facilities before and after storage
2. Regular inspection of storage condition.

Types of Storage

Storage systems maybe classified according to storage capacity, handling method or container structural materials.

Farm House Storage

1. Sacks
2. Wooden boxes
3. Bamboo baskets
4. Cans
5. Drums

Granary (Sack or bulk handling)

1. Wooden
2. Bamboo
3. Sheet metal
4. Concrete

Warehousing (Indoor)

1. Sacks in piles

2. Uncovered bulk container
3. Flat store for bulk storage

Silos (Outdoor)

1. Metal sheet (flat structures, bulk handling)
2. Concrete (tall structures, bulk handling)
3. Clay-straw silos

Grain Aeration

The storage of foodgrains is normally done for extended periods in order to maintain a uniform supply of food for consumption, for the domestic and export market and to provide a buffer stock for contingencies such as drought floods and war. Millers, traders and other private entrepreneurs adopt it to speculate on good price. Like any post production operation, losses in storage are considered significant. And these are attributed mainly to spillage, attacks by insects, mites, rodents, moulds and dry matter loss due to respiration.

In Southeast Asia, for instance, it is reported that grain losses in storage ranges from 5% to 15%. Insect infestation accounts for the single largest component of these losses. The storability of grains is affected by their temperature and moisture content. Either or both factors must be reduced to ensure a conducive environment for storage. This could be done through aeration.

Aeration involves moving a relatively low volume of air through the grain bulk to control grain temperature. The process was found to reduce the risk of damage or spoilage of grains. Aeration is normally done using ambient air. The cooling process is termed ad ambient air aeration or natural aeration. Otherwise, dehumidified air is used as in other countries where the ambient air may have too high a heat content or enthalpy to effect sufficient cooling.

Benefits of Aeration

Temperature control

There are two general objectives of aeration. These are: a) to maintain a uniform temperature in the grain bulk; and b) to keep that temperature to a low level as practical.

Like most stored products, grain is a poor conductor of heat. As such, heat does not dissipate or escape easily or quickly in some portions of the bulk. Non-uniformity of temperature keeps warm spots to remain warm and if high temperature differences exist in the bin as induced by solar radiation, air convection currents are generated causing moisture migration. In this phenomenon, the convected air picks up moisture from warmer grain and transfers this moisture to the cooler grain where condensation of moisture would likely take place. This results in grain damage which is attributed to moulding, caking, rotting, and sprouting. This is considered critical in areas where large seasonal changes in temperature exist.

The respiration of insects, moulds and the grain itself also creates localised heating in the grain bulk which could also raise the temperature of the region they occupied. This center of insect activity is known as the "hot spot". The hotspot expands in size because insects migrate because of its high temperature and create identical conditions along side through further respiration. The water produced by respiration tends to rise in the warm air of the hot spot and condenses in the cold grain. Both phenomena can cause damage and loss due to mould infestation.

Maintaining a low temperature in the grain bulk can deter the development and growth of fungi and could also inhibit insect infestation. Generally, at low temperature, the rate of multiplication of insects is very low. For instance, the saw-toothed grain beetle will not breed if the temperature is below 18°C. The grain weevil, on the other hand, can breed at temperatures as low as 13°C. Other studies have shown that cooling the grain to 63° F (17°C) or below prevents granivorous insects from completing their life cycle quickly enough to cause significant build up and damage to grain. Furthermore problems on insects are limited to temperatures in the approximate range of 15-40°C.

In tropical climates, however, storage temperatures are normally in the range of 20-30°C. Hence, cooling or heating the commodity to temperatures near or beyond the limiting temperatures gives a measure of control over insect development. Low temperature in the grain bulk can significantly reduce the amount of pesticide required to give longterm protection from insects. Also, the rate of decay of pesticides is independent of the initial condition of grain. This means that the final concentration of pesticides on aerated grain after a given storage period is about the same, whether the grain be initially warm and wet or cool and dry.

Microbial growth is enhanced with increase in temperature. Reports have shown that an increase in growth of about 2.5-4.5 fold can be expected with 10°C increase in temperature if the temperature does not exceed the optimum for fungi. The optimum range for a variety of storage fungi is about 23 to 40°C.

Other Uses of Aeration

1. Removing odors from grain—unnecessary odors such as those associated with the use of chemical

preservatives, mouldy or sour odors, etc. can be removed or reduced in intensity by aeration.

2. Equalising grain moisture—in bulk storage, there is greater chance of storing lots of batches of grain with varying moisture contents. This moisture variation can be equalised through aeration.
3. Fumigant application—grains in deep bins and silos can be effectively applied with fumigant through the aeration system.
4. Holding moist grain—newly harvested grain can be stored in receiving bins in short periods without appreciable damage using aeration to provide cooling and dessipating heat caused by respiration. This method of holding wet grain is important to ease out the varitation of grain received during peak harvest. Studies have shown that corn can be kept at moisture ranging from 24% to 26% if it is cooled quickly to below 50°F.
5. Removing dryer heat—sometimes called dryeration or bin cooling. Aeration is applied in tempering bins after every pass in continuous flow multistage drying.

Components of an Aeration system

The components of aeration system basically consist of the following:

1. Fan—of sufficient capacity to supply the airflow requirement.
2. Supply-duct—conveyance for the aeration air leading from the fan to a perforated duct placed in the store or plenum beneath the bin.
3. Provision for airflow—into or out of the air space over the grain surface.

If recirculatory fumigation is to be carried out, a return duct is also included. Except for this return duct, the above basic components are found for both the bulk and bagged grain aeration systems. Figure 3 shows the typical bulk storage and bagged grain storage system, respectively, with the principal components of an aeration system.

Aeration Principle and Systems Design

The temperature of the grain stored in bulk is reduced by allowing cool air to flow through the warm grain mass. The grain is cooled by displacing the warm air and by contact cooling. Ideally, the moisture content does not significantly change as the relative humidity of the aeration air is in equilibrium with the moisture content of the grain. When the cooling air has a low relative humidity, however, drying may take place. If this condition persists, then the grain may be cooled to below the incoming air temperature due to the evaporation from the surface of the grains. Conversely, the delivery of air above the equilibrium relative humidity of the grain at a given moisture content will result in rewetting the grain. Furthermore, climate and moisture content determine the requirement for aeration.

There are three principal considerations in the design of aeration systems. These are 1) airflow rate, 2) fan selection and 3) air distribution. Automatic controls which are now widely used mayle part of the system.

Airflow rate—This is the volume of air desired to maintain uniform conditions in the stored bulk and to remove the generated heat and water. The recommended rate depends on the purpose of aeration, the type of grain being aerated, the size and type of storage structure, and climatic conditions. In the United States, an airflow rate of 2 liters per second per ton is widely used for shelled corn and soybean stored in farm bins and in flat types of storage. For wheat and other smaller seed grain, one (1) liter per second per ton is more common. A range of 1 to 0.5 liters per second per ton is widely used especially when cost limitations for fan power is considered.

Fan selection – The selection of fan is normally based on the airflow rate used for a particular grain, the kind of grain handled and the grain depth. These factors determines the resistance of grain to airflow and the static pressures against which the fan must deliver the required airflow.

Two types of fan are used for aeration. These are the centrifugal and the axial flow fan. generally, the axial flow fan will deliver more air than centrifugal fans at a static pressures up to about 4 inches of water (1,000 Pa). For higher static pressures, the centrifugal fans are recommended.

Air distribution – This includes the ductings, false floors, etc. which are used to move the air to the desired points. The proper sizing of the ducts, the sizing and spacing of the openings in the ducts to let the air move between the duct and aerated grain, the layout of the duct system are important to maintain the entering air at an acceptable velocity and provide uniform airflow through the grain. Teter has outlined a detailed procedure in the design of aeration in bulk storage.

Aeration System Operation

In controlling temperature, continuous ventillation is necessary during the early stages unless air temperatures are excessive. But for economic considerations, fan operation is selected only at a particular time of the day when the ambient relative humidity is lower or equal to the equilibrium moisture content of the grain. The use of humidity controllers which is interlocked with the fan controls provide convenience in the operation of the fan. Generally, the grain temperature is reduced to below 15°C as soon as possible. Once the desired temperature is reached, only minimal ventillation is required. In this aspect, temperature sensors are likewise convenient to monitor and check the temperature in the bulk.

9

Safety in Insecticide Usage

The term "safety" means different things to different people and its meaning differs mostly with each situation in which it is used; at best it is a very relative term. For purposes of those of us who deal with insecticidal chemicals, it involves the concept of hazard or risk, i.e. the likelihood of some adverse effect resulting from certain sets of circumstances, including the approved use of a chemical or any possible misuse.

Hazard in the use of pesticides may be defined as the likelihood that injury could result from the use of or exposure to the product, Often, hazard and toxicity are considered by many people to mean one and the same. Toxicity measures the actual harm which results if the product is absorbed while hazard takes into consideration the likelihood of contact or absorption. In many instances toxic substances are handled for long periods with out any harm simply because precautions have been taken to prevent contact or absorption, hence hazard become negligible.

Handling and Usage of Insecticide

In order to better appreciate safety of insecticide handling and usage, it is important to consider the following hazards and take the necessary precautions:

Acute Hazards

Oral intake

1. Eating the substance or contaminated food.
2. Drinking the substance or contaminated food.
3. Eating or drinking from contaminated utensils.
4. Eating or handling food with contaminated hands.
5. Blowing or sucking to clear a blockage in equipment.

Dermal absorption

1. Handling concentrates without protection.
2. Splashing concentrates onto skin.
3. Spillage of concentrates.
4. Contaminated clothing, tools, or work places.
5. Lack of protective clothing.
6. Carelessness in mixing and spraying

Inhalation-There is a risk of absorption through the lungs

1. During fumigation
2. Due to insufficient ventilation in work area.
3. During aeration of treated grains
4. During mixing and spraying contact insecticides

Chronic hazards

These can result from faulty practices repeated over extended periods:

1. Familiarity and failure to observe precaution
2. Poor personal hygiene

3. Poor ventilation
4. Inadequate protective clothing
5. Faulty equipment
6. Repeated exposure to spray or dust.

These are mainly direct hazards affecting men. Hazards to the environment is an important consideration particularly its negative ecological impact. For our purpose we will concentrate on the dangers to man, the applicator and consumers. It does not necessarily mean that insecticides are bad and should not be used—far from that. Realising these hapards, let us now consider the corresponding precautionary measures to counteract these risks.

Mammalian Toxicity

Toxicological studies are conducted to determine the threshold limit of a chemical which an animal or human is capable of handling without significant biological effects. The usual beginning in any toxicological evaluation is the assessment of the acute toxicity i.e., the effects of a single dosage of the chemical. The general technique is the determination of the LD50. The compound is administered on a weight/weight basis in a suitable solvent or suspension system. This is evaluated by acute tests, orally (AO) or dermally (AD), chronic oral tests (CO), vapor toxicity tests (VA) and chronic vapor tests (VC) or inhalation tests (IT).

Insecticides can be classified according to their toxicity based on the LD50 values:

1. High toxic

   AO LD_{50} = 0-50 mg/kg

   AD LD_{50} = 0-200 mg/kg

IT LC_{50} = 0-2000 ug/l

Danger, skull and crossbones and poison on label.

2. Moderately toxic

   AO LD_{50} = 51-500 mg/kg

   AD LD_{50} = 201-2000 mg/kg

   IT LC_{50} = 2,0001-20,000 ug/l

Warning on label.

3. Slightly toxic

   AO LD_{50} = 501-5000 mg/kg

   AO LD_{50} = 2,000-20,000 mg/kg

   IT LC_{50} = more than 20,000 ug/l

4. Relatively nontoxic

   AO LD_{50} = 5000 + mg/kg

   AD LD_{50} = 20,000 + mg/kg

   AD LC_{50} = of some of the common insecticides

Generally, the insecticides used in stored product treatment are of low mammalian toxicity (Table 1) and are in a formulation that is likely to be effective against the species involved, persistent for the required period of time under given storage conditions and will not alter the flavor, color of the stored commodity.

Insecticide Residues

The Joint FAD/WHO Codex Alimantarius Commission was established to implement the Joint FAD/WHO Food Standards Programme. The purpose of the Programme is to elaborate international standards for food aimed at protecting the health of the consumer, ensuring fair practions in the food trade, and by their acceptance by governments resulting in the harmonisation of national food legislations, thereby facilitating international trade.

The food standards incorporate provisions in respect to food hygiene, food additives, pesticide residues, other contaminants and methods of analyis and sampling.

Table 1. LD_{50} of some insecticides used for stored product protection'

Insecticide	*LD_{50}, Acute Oral mg/kg body wt. rat)*
Malathion	885 - 2800
Pirimiphos methyl	2050 - 2300
Chlorpyrifos methyl	041 - 2140
Tetrachlorvinphos	4000 - 5000
Bromophos	3750 - 7700
Dichlorvos	25 - 170
Diazinon	66 - 600
Fenitrothion	250 - 670
Lindane	76 - 200
DDT	87 - 500
Methoxychlor	5000- 7000
Carbaryl	307 - 986
Alletrin	680 - 1000
Pyrethrine	200 - 2600

The Codex has a committee which establishes maximum limits for pesticide residues in specific food item or group of food.

Guidelines for Safe Pesticide Use

Although the use of insecticides has its undesirable side effects, its use must go on until a better control method has been found. Thus, there is a need for precautionary measures in its use so that the harmful effects can be minimised.

Precautionary Measures

1. Never apply more or less than what the pesticide label specified time of application and rates at the specified time of application and follow safety precautions.
2. Store pesticides under lock and key, away from children, pets, food and anything or anybody who does not understand how dangerous they can be.
3. Never leave pesticides unlabeled or misleadingly labeled. Better still, keep them in their original containers, whenever practicable.
4. Do not smoke or eat while handling pesticides. Protective clothing, gloves and masks are desirable when one works closely with them. Discard contaminated clothing and empty packages and always wash thoroughly after handling such toxic chemicals.
5. Never re-use pesticide containers for purposes such as for storing food or water. No one is ever sure that residues will not be there.
6. All poison baits should be clearly marked as dangerous. When baits containing acute poisons are laid, everyone should be excluded from the area; and when treatment is finished all uneaten baits should be collected and burnt along with any rodent corpses found during and after treatment.
7. If symptoms of poisoning (nausea, rashes, vomiting), occur, stop working and run, don't walk, to the nearest antidote. The life you save may be your own.
8. In general, repeated and prolonged contact with any insecticide is to be avoided especially if it is in

the form of dusts, mists, vapours, or concentrated formulations.

Curative Measures

1. *Curative measures*—Spillage, breakages and other such accident cannot be entirely ruled out in insecticide use. The following therefore are geared towards minimising poisonous effects:
    a) Spilled materials are immediately wiped with rags which are then buried. Do not burn.
    b) Thoroughly wash with soap and water skin that has been in contact with pesticides. Contaminated clothing should be washed thorouhly before use or burned depending on the degree of contamination.
    c) If pesticide has been swallowed, induce vomiting and immediately seek medical attention.
    d) In cases of dizziness or nausea in the use of pesticides,. stop work immediately and get medical attention.

All pesticides must be considered in some way potentially harmful to man and should only be used by trained persons.

In particular, concentrated insecticides, rodenticides and fumigants must be handled with extreme caution. Washing water must be available to operators.

Protective clothing should be available. Rubber gauntlet gloves should be worn when handling concentrated insecticides. Face masks must always be worn when insecticide dusts and sprays are mixed or being applied. They should also be worn when rodenticide dusts or concentrates are being used.

For fumigation involving methyl bromide, full face respirators with the appropriate canisters must be used. For phosphine fumigations, respirators with appropriate canisters must be available for all but the smallest trements. In all but the smallest fumigation treatments, two operators should be present with one designated as in-charge.

When mixing concentrated insecticides with water or other diluent, a spoon or other implement must be used. Never use the hands to pick-up concentrated insecticide. During this operation, especially if the store room is small, always leave the door open to prevent a build-up of dust or fumes. Lids should be replaced on concentrate concentrate containers immediately after use.

Care must be taken to ensure that correct application of insecticides is undertaken particularly where insecticides are added directly to food commodities.

Always read the lable on the container and follow the manufactures instructions.

Application equipment for sprays and fumigants must be checked before the start of treatment.

Rodenticides must be placed in locations that are not accessible to domestic or farm animals. Use bait boxes where where possible.

All buildings or stacks of products being fumigated must be clearly labelled. Personnel should not continue to work in the building whilst a fumigation is being carried out and should not resume work there until tests have shown it is safe to do so.

All pesticides should be stored in clearly labelled containers in a safe place under lock and key. Storage should not be in the main warehouse where food is stored, but in a separately designated room. Pesticides should not be stored in or adjacent to living quarters.

Storekeepers in charge of pesticides must be properly trained. They must not eat, drink or smoke in pesticide stores. All pesticides must be checked into and out of stores and a record with full details kept. Large stocks of pesticides are best avoided since they can deteriorate quickly at high temperatures and humidities. Efforts should be made to ensure that sufficient pesticides for one season only are stored wherever possible.

Recommendations for Good Agricultural Practice in the Use of Pesticides

These guidelines are intended for use by administrators, specialists and advisory agencies. No attempt will be made to elaborate recommendations for the use of individual pesticides and their formulations. It should be emphasised that due to differences in pests, -pest population, commodities, climate and geographical location, it is not possible to propose universal recommendations for the use of specific pesticides against specific pests. It should also be remembered that not all countries have uniform capabilities for the development and implementation of detailed guidelines for the use of pesticides. Therefore, only general guideline practices can be suggested for the use of pesticides, and the details of use will require development of information within the Member countries. The important issue is, however, to try and eliminate disturbing side-effects of certain pesticides or certain applications.

A first basis for general guidelines is provided by the Codex definition of the concept of "good agricultural practice in the use of pesticides", which is as follows: "Good agricultural practice in the use of pesticides is the officially recommended or authorised usage of pesticides under practical conditions at any stage of production, storage, transport, distribution and processing of food, agricultural commodities and animal feed, bearing in mind the variations in requirements within and between regions, which takes into account the minimum quantities necessary to achieve adequate control applied in a manner so as to leave a residue which is the smallest amount practicable and which is toxicologically acceptable".

1. Noting that the occurrence of unintentional residues in a number of food items and animal feedstuffs is partly a result of environmental contamination, the Meeting recommends that efforts be made to discover the sources of such contamination and, where possible, to eliminate them, in order to reduce the background level of pesticide residues.
2. In view of the concern over the extent of the use of certain persistent pesticides, the Meeting recommends that they be replaced, wherever possible, by pesticides, the residues of which are less undesirable toxicologically.

The emphasis in these recommendations is clearly on the desirability of making a conscious effort to replace certain pesticides by alternative chemicals, which are preferable from the food or environmental health point of view. The importance of reducing residues in animal feed has again been emphasised by a recommendation of the 1969 Joint Meeting.

"Because some compounds currently in use as seed protectants are highly toxic to humans, and their uses can result in the occurrence of unintentional residues, the need to develop safer substitutes was emphasised. In the meantime, every effort should be made to reduce the contamination of commercial grain and animal feeds to the minimum and to undertake surveys to ensure that directions are being observed by farmers and others concerned with the handling of treated seeds".

Pesticide Situation

For the foreseeable future agricultural pesticides will continue to be required in the production, transportation and storage of food, feed and fibre. However, most pesticides are derived from, or produced through the use of non-renewable world resources. Wastage in their use should be avoided as part of an overall global effort to conserve these resources, maximise their effectiveness and minimise unnecessary environmental pollution.

Apart from occupational accidents and gross misuse, the regular use of pesticides has resulted in a number of undesirable side-effects, partly attributable to occasional indiscriminate applications, partly due to undoreseen biological effects.

This situation has prompted developments which can be briefly described as follows:

(a) research on alternative methods of pest control, and introduction of these wherever technically possible and economically feasible;

(b) introduction of alternative pesticides with greater safety and suitability;

(c) more adequate legislation or other administrative provisions, coupled with better extension and education facilities in the field of pesticide usage.

At present alternative methods of pest control are not available to the extent that they can be applied on a broad scale as full replacement of pesticides, but they offer possibilities in specific cases, either alone or in combination with selected or selective pesticides. Research on alternative methods of pest control to be applied in combination with selected or selective pesticides should, therefore, be intensified, and in the meantime emphasis should be placed on a system of supervised control, which aims at a judicious use of pesticides coupled with assessment of economic threshold levels and forecasting systems.

Pesticide legislation or other effective control systems should be introduced and implemented in those countries where they have not yet been established; in other countries with existing legislation or administrative controls, these may need to be intensified in order to strike a better balance of benefits to agriculture against risks to humans, environment and food.

The extent to which regulations are introduced should, on the other hand, not go beyond reasonable and acceptable limits, in order to ensure that plant health and other pest control requirements as well as availability of adequate pesticides are not jeopardised. Regulations should be accompanied by enlightening and meaningful educational programmes on pesticide use and safety.

Legislation on pesticides deals mainly with two fundamental aspects, which are distinctly different though interrelated, i.e.:

(a) use pattern and handling regulations, pertaining to registration and approval on the basis of criteria for effecacy and for side-effects;

(b) residue regulations, pertaining to the establishment of maximum limits of pesticide residues in food and feed on the basis of "good agricultural practice".

Pesticide residues occur in agricultural commodities as a result of (a) intentional use of pesticides for protection of growing crops or stored products or of animals; (b) unintentional exposure to pesticides such as would occur in crops grown in soil treated previously or contaminated by foliar treatments of other crops grown earlier in the rotation; (c) unintentional accumulation in animals from the ingestion of feeds containing pesticide residues; and (d) contamination of crops or animals exposed to chemicals in the environment.

It has to be recognised that differences exist both between and within countries as to pest incidence, pest control conditions, and types of crops, which may be reflected in the different pesticide use patterns, pesticide demands and requirements regarding maximum residue limits.

The prime target of the Codex Alimentarius Commission is to reach agreement on international maximum limits for pesticide residues in food, in order to avoid trade barriers and to secure good agricultural practice under widely varying conditions. As maximum residue limits, use pattern and good agricultural practice are interrelated, the Codex Committee on Pesticide Residues has also undertaken to formulate these guidelines for the use of pesticides.

These guidelines indicate principles for the use of pesticides in agriculture, and in the harvesting, marketing, transport and storage of foodstuffs. Taking into account the attainment of the desired degree of control of pests at an economic cost and with a minimum of danger to operators, agricultural workers, consumers, beneficial animals and the environment, the following represents a list of goals which should be aimed at in good practice in the use of pesticides for the above mentioned purposes. It should be understood that the information presented in the guidelines is not intended as a substitute for actual supervised trials under the growing conditions of the area involved.

General

(a) If pesticides reach humans or animals through different routes and thus give rise to additional body loads, the use patterns may have to be adjusted, and if necessary, priority should be given to those uses which are indispensable and for which no adequate alternatives are available.

(b) Maximum residue limits established for products for human consumption are not necessarily acceptable for the same product when this is destined for animal consumption and vice versa, and in such cases this should be indicated as far as possible.

(c) In view of the necessity of preserving a balance between cost, productivity, quality and freedom from residues, the concept of good agricultural practice in relation to pesticide residues embraces all interrelated and essential factors and functions, which ensure that the pests will be controlled effectively, leaving residues that are the smallest amounts practicable and that are toxicologically acceptable.

(d) Therefore, pest control treatments should only be made when necessary. The requirements for pest control should first be established, followed by the application of the preferred method of control.

Choice of Pesticide

(a) All pesticides which are used should be authorised

by appropriate authorities in the country of use. They should only be marketed with labels indicating recommended or approved uses, times, methods and rates of application, and safety precautions for the uses. Such recommended methods of application, should be based on supervised trials and other experimental work, and should take into account such variations in climate, in crop husbandry, and in incidence of pests as may occur under practical conditions from time to time in the various places in which the pesticide may be used.

(b) Bearing in mind the actual conditions under which the pesticide will be used, the pesticide should be adequately safe to humans and the environment, and at the same time provide adequate pest control.

(c) Where a choice of pesticides is possible, the cost and effectiveness of available pesticides should be weighed against the risks involved, and those which show a more favourable benefit-risk ratio for the particular purpose in question should be preferred.

(d) When pest control is required in the early growing stage of the crop, a pesticide may be needed which has an adequate and acceptable degree of persistence, in order to avoid repeated applications of non-persistent pesticides.

(e) Where plant quarantine and/or phytosanitary requirements make is necessary to apply pesticides close to or after harvest, those which have a short persistence should be preferred.

(f) The agricultural use of persistent and/or cumulative pesticides on crops for human consumption should be restricted as much as much as possible, and be limited to the control of pests, weeds and diseases for which at present no suitable alternative chemicals are available.

(g) As a general rule, persistent and/or cumulative pesticides should not be used on fodder crops and not be applied directly to animals for veterinary purposes.

(h) Where post-harvest treatments are required, pesticides which leave residues that are the smallest amounts practicable and that are toxicologically acceptable, do not interact with the food commodity, and/or are readily removed during storage, preparation or cooking, should be preferred.

(i) With respect to post-harvest treatment of stored products it is recommended no to use persistent and cumulative pesticides as direct admixture.

(j) The application of adequately durable pesticides to the exterior of packing material for stored products is acceptable, but the use of highly persistent and cumulative pesticides should be avoided as much as possible.

Choice of Formulation

(a) Formulations which combine maximum efficiency of the pesticide with minimum risk should be preferred.

(b) Supplementary adjuvants should be used only if their effects, including toxicological effects, are known and where their use produces a significant improvement in performance.

(c) In general, the use of combined pesticide/ fertiliser formulations should be avoided, However, such practices are recommended by local authorities when they are considered beneficial.

Dosage

(a) The quantity of pesticide applied should not be greater than the minimum required to achieve the desired degree of control.

(b) The number of treatments should be determined by the desired degree of control and by the severity of pest conditions.

Application

(a) The method of application should be selected to ensure optimum pest control with the minimum contamination of the crop and the environment.

(b) Indirect treatment can, in some cases, be used to supplement or replace direct application to food crops.

(c) Application equipment should at all times be maintained and used according to the makers' instructions.

Timing of treatment

(a) Treatment should preferably be carried out when the pests are at the most vulnerable stage of development, and when climatic conditions and cultural practices will ensure that the optimum effect can be attained from the treatment. In some instances, however, action may be necessary immediately following detection of the pest species.

(b) The interval between last application and harvest should be as long as possible in order to permit the greatest reduction in pesticide residues, bearing in mind the pest incidence, the degree of control required for a maximum utilisation of the commodity, and the vulnerability of the treated crop immediately preharvest. To this end official pre-harvest intervals should be established and adhered to.

Post-treatment practice

(a) Crop rotation should be adjusted in such a manner that unintentional residues in the edible parts of the crop, as a result of previous treatments, will be minimal, particularly if the crop may be used as animal feed, and accumulation in the animal body may lead to undue residues in food products of animal origin.

(b) Seed-grain, treated with pesticides at dosages to provide long-term protection in the soil, must, under no circumstanes, be mixed with commodities destined for human or animal consumption. Sufficient safeguards ought to be provided which would minimise the accident risk of such practices.

(c) Where grain intended for consumption must be protected in storage, only compounds with low chronic toxicity and/or shourt persistence should be used.

(d) In storage practice, the selection of the pesticide for treatment of empty warehouses or ship holds and the subsequent storage arrangements should be such that there is a minimum risk of contaminating feed or food products.

Problems in Use of Insecticides

Resistance Development

Resistance is an increase in tolerance status of a particular pest species to a toxic substance. The onset of resistance as demonstrated by Parkin, assuming that population has the genetic potential, is dependent on

1. The life cycle must be relatively short.
2. Insecticide stress must be exerting some selection pressure on successive generations.
3. Stressed populations must be relatively free from immigration of susceptibles.

The problem of resistance development is accentuated in bulk grain by the relatively constant and protected microclimate, due to the insulating action of both the storage structure and the grain itself. Insect populations are therefore not subjected to any violent fluctuations in seasonal conditions and breeding is possible throughout the year. There exists a dynamic variation in tolerance among individuals of any given population, and resistance is therefore initiated by continuous selection of the toxicant of the more resistant individuals in that population.

Resistance is due to the selection of hereditary factors and not an adaptation of individual insects to the insecticide. Genes are selected that, alone or in combination, alter the physiology or biochemistry in such a way that progressively higher amounts of insecticide are required to kill them, even up to a point where an insect can tolerate any given dose. The point to remember is resistance will result directly from selection pressure and from this it follows that:

1. Extent of selection is dependent on numbers of insects available for selection, and hence any reduction in the numbers available reduces the genetics variability and therefore the probability of resistance develoment. It is much dependent on strict hygiene and cleanliness and good ware housekeeping to reduce the breeding potential and buildup of insect populations.
2. Controls applied should aim at complete mortality to prevent selection occurring, i.e., the full dose should be available to every individual and assumes that the prescribed rates and concentration is sufficient to kill the most tolerant members of any normal population. Fumigation has been cited as one of the reasons why stored product insects were slow in developing resistance compared to household pests, and to pests of medical and veterinary importance, because mortalities approximating 100% can normally be obtained.

Fumigation is recommended in case of control breakdown following use of residual insecticides. Why? Because control breakdown maybe the first indication of tolerant members of that endemic population surviving the recommended treatment. His therefore essential to kill them so as to maintain the use and efficacy of the small number of residual contact insecticides registered for use as grain protectants

Mechanism of Resistance

Insect resistand was first discovered in DDTresistant houseflies, Musca domestica in 1946 and has since become one of the major constraints/concerns in insect pest control. The most widely used stored-product insecticide during the 1960's to early 70's was the organophosphate Malathion 0,0-dimethyl dithiophosphate). Malathion was introduced because of its low toxicity to mammals and birds, its low phytotoxicity and its rapid breakdown rate.

Table 1. General rate expressions for 12 protectants plotted to the equation

$$\log t_{1/2} = \log(t_{1/2})_0 - B(T_2 - 30) - \log RH/50$$

where $t_{1/2}$ is the half-life on grain of temperature, T_2 in degrees Celsius and percentage relative humidity RH, and $(t_{1/2})_0$ is a reference half life on grain of 30°C and 50% humidity.

Pretectant	$t_{1/2}$ *(weeks)*	*B (degrees^{-1})*
Fenitrothion	14	.036
Bioresmethrin	24	.033
Bioresmethrin	38	.031
d-Fenothrin	38	.029
d-Fenothrin	40	.029
Pyrethrum	55	.022
Methacrifos	8	.055
Malathion	12	.05
Chlorpyrifos-methyl	19	.04
Carbaryl	21	.031
Pirimiphos-methyl	70	small
Pyrethrum	34	-

In Australia, Malathion was applied at the beginning of the storage period when the risk of infestation was highest with normally only one application is used. Work by Minett and Williams using 150 9 masses of wheat at 25°C and 70% RH showed that after 10 weeks about 50% of the malathion still remained after an initial application of 12 ppm. Table 1 depicts the half-life of some grain protectants at the reference temperature and relative humidity of 30°C and 50% RH respectively, which are typical of receival conditions for grain in Australia.

Malathion acts as a contact poison which is absorbed through the cuticle. Once inside the body it is oxidised to malaoxon $I(CH_3O)_2P(0)SCH\ (CO.OC_2H_5)CH_2CO.OC_2H_5I$, which then acts as a powerful cholinesterase inhibitor. This process of conversion of malathion is known as activation, resulting in a disruption of nerve activity due to acetylcholine accumulation at the nerve endings which results in excitability, muscular tremors and finally paralysis.

Acetylcholinesterase (AChe) is the main site of action for the organo-phosphorous and carbamate group of insecticides. Messages are transmitted across the nerve synapse by the transmitter acetylcholine, and once the impulse is passed acetylcholinesterase acts on acetylcholine to return it to its active ionic state. The organo-phosphorous insecticides of the -thionate configuration must be oxidised from the P = S to the P = 0 nucleus before they become active, since the P = 0 nucleus has a strong electronic pull on the hydroxyl group of acetylcholinesterase, thus bonding with it and inactivating it once the insecticide penetrates the synapse.

Resistance can be due to various changes, either singly or due to their interaction such as:

1. reduced uptake or penetration of the insecticide
2. increased detoxification, storage and excretion
3. reduced transportation to the site of action, or
4. reduced action by an alteration in the site of action

In most cases of insecticide resistance, increased detoxification and altered site of action appear to be responsible. This can be brought about by enzymatic changes caused by mutants of single genes. Single gene alteration causing a change in the site of action has been demonstrated in spidermites, cattle ticks, leaf hoppers and houseflies where resistance has developed due to an altered cholinesterase displaying a much reduced rate of reaction with the inhibitors. High degrees of resistance due to single gene alteration resulting in the production of more efficient detoxifying enzymes is known in DDT-dehydrochlovinase and carboxyesterase strains of houseflies degrading DDT and malation respectively. Both resistances are due to the altered properties of the enzyme as well as greater quantities of enzyme produced.

High levels of resistance can also develop through the interaction of more than the resistance gene which may result in decreased penetration as well as detoxification of the insecticide. Resistance to a single insecticide can be due to different causes in various strains of the same species. Malaoxon can be degraded by a carboxyesterase as well as by a microsomal mixed function oxidase in the housefly. They can be present either singly or together in the same strain eliciting different behavioral responses with applied synergists, such as triphenyl phosphate (TPP).

Malathion is unique in that two distinct metabolic resistance mechanisms are possible. Malathion non-specific resistance (*Mal. nonspec R*) is a broad resistance to most organophosphates (i.e. cross resistance). This type of resistance is induced by selection or exposure to any organophosphate compound. Malathion specific resistance (*Mal. spec. R*) is restricted almost exclusively to malathion. This type of resistance is due to higher levels of a carboxyesterase in the resistant strain. Although first demonstrated in the mosquito, Culex tarsalis,, it has since been demonstrated in a number of other insects. The carboxyesterase attacks the $CO.OC_2H_5$ group of the malathion molecule and thus prevents activation.

Genetic studies have shown that high carboxyesterase levels and malation specific resistance are inseparable connected. Increased mixed function oxidase activity is a very important mechanism since it affects insecticides from practically all chemical groups, whereby crossresistance has been conferred in insect growth regulators with the oxidases being amongst the most likely candidates. Through selective pressures from the use of IGR analogues, high levels of resistance may develop, but Amos et. al., detected no crossresistance to either methoprene or hydroprene in malathion susceptible or resistant strains of *Tribolium castaneum* or *T. confusum*.

There is also evidence that the different forms of the detoxifying enzymes and the aberrant mutants of the cholinesterase are produced by alleles of the structural gene. The structural gene continues to produce a functional protein while the duplicate is free to evolve without subjecting the population to long periods of selective disadvantage. They found resistant strains contained more carboxylesterase enzyme than non-resistant strains and that enzyme activity increased exponentially with insecticide resistance brought about by gene doublings. Malathion and parathion resistance in different strains of the housefly Musca domestica as well as altered cholinesterase in resistant spider mites Tetranychus urticae is thought to be alleles of structural genes.

Truly sublethal exposures to applied insecticides do not induce resistance in known susceptible strains of *M.*

domestica or increase the levels of DDT dehydrochlorinase (DDT-ase) as demonstrated by Moorfield and Brown. Rather, the cumulative effect of daily sublethal amounts of DDT, gamma BHC, dieldrin or diazinon renders the housefly more susceptible to exposures at the discriminating dose. Bond and Upitis with sub-lethal treatments of phosphine, found cumulative effects which differed with species when repeatedly applied to insects.

For example *Tribolium consusum* adults recovered from a sublethal treatment in 10 days while *Sitophilus granaries* adults required 40 days although the tolerance level regained was less than in untreated individuals. Tenebroides mauritanicus larvae showed a varying response after an initial dose that achieved no mortality in the first two weeks; some individuals died slowly, some remained permanently as larvae, some developed into pupae but subsequently died, while some adults that emerged were physically deformed and some appeared morpholigically normal.

These observations on increased susceptibility of insects exposed to sub-lethal doses of phosphine maybe important in terms of economic control if for example, survivors from a treatment where incomplete control is obtained, are more responsive to subsequent treatments applied soon after the first. The normal criterion for effectiveness of a treatment is usually mortality, but various sterilising effects that inhibit further development of insect populations becomes important in considering combined control strategies.

The mechanism of resistance is by no means clearcut, but certain established criteria enable pest control operators to implement counter measures where suspect resistance occurs.

1. Development of resistance to one particular pesticide confers on insect populations a measure of resistance to related pesticides, i.e., in same chemical group. Low level resistance to other unrelated series of pesticides may also be developed concurrently.
2. Pests in the field maybe exposed to a number of different types of pesticides and, together with introduction of genetic material from other migrating populations or by cross-infestation with strains carrying other types of resistance may result in a complex of unrelated resistances being present in the same strain.
3. Where resistance to a particular series of pesticides has been developed, these tolerant populations will develop high levels of resistance to unrelated chemical groups with greater facility than would populations previously unexposed to pesticides.
4. Resistance in field populations maybe lost on segregation of these populations from selection pressure or exposure to pesticides. Rate of loss is dependent on the genetic constitution of the pest and the mode of inheritance of resistance, the intensity of selection for resistance and dilution of the population by individuals of normal suseceptibility from unexposed populations.

When resistance has been observed, the pesticide in use should be replaced by one of an unrelated chemical groups and attention given to its efficient utilisation. With residual pesticides the first indication of resistance is a reduction in the time that residual deposits remain effective; thus a grain protectant that inintially gave 6 months protection may subsequently not control insect populations although residue analyses may indicate no change in the normal decay pattern of the pesticide on the commodity. Early warning indications such as this often go unheeded through mismanagement and it is not until obvious signs of resistance such as large scale buildup of pest populations in or on treated commodities that the problem is recognised. Regular monitoring of the tolerance status is a valuable means of early detection of resistance and can provide unequivocal evidence of the contribution of resistance to control failures.

Breakdown in control measures other than those initiated by insecticide resistance are as follows:

1. Use of pesticides that have deteriorated in storage.
2. Incomplete coverage or application due to faulty equipment or treatment:
   - blocked nozzles leading to underdosing.
   - incorrect calculations on flow rates overdosing.
3. Use of unstable preparations: rapid breakdown or disregard of comparability warning on labels.
4. Variation of pest susceptibility with temperature. Dosage levels maybe effective at normal temperature but inadequate in extremely hot or cold weather. Also affects stability, reaction rates of insects and rates of diffusion with chemicals.
5. Also, under conditions of extremely hot weather and light, pesticides maybe decomposed or otherwise lost at abnormally high rates.

Resistance in stored products has advanced from 3 species in 1960 to approximately 17 species of Coleoptera and Lepidoptera in 1979. Resistance to fumigants detected in the field is of major concern because of the global dependence on fumigation both as routine disinfestation and for combating insecticide resistance strains. From the 1972-73 FAO global survey of tolerance status of the major pests of grain, methyl bromide resistance was detected in 23 countries and phosphine resistance in 35 countries, where resistance factors were generally low and in most instances would not have resulted in breakdown of pest control.

The emergence of strains of different species may not be a practical problem as yet in the field, but the widespread occurrence of strains surviving discriminating doses which would normally prove fatal to susceptible strains and the case at which this level of tolerance can be increased under laboratory selection, poses a serious threat to international trade. Most cereal products moving in international trade is almost exclusively in bulk ships and the infestation either originates from the exporting country or, from cross-infestation from residues in previous cargoes or other infested cargoes on board.

Malathion resistance was first detected in *Tribolium castaneum* in the early 1960's and now can almost be considered to be a normal attribute of a *T. castaneum* population. The frequency of resistance in other species is still comparatively low and although a much later detected resistance in these species indicates genotypes which were not as conducive to development of resistance.

The movement of resistant strains in trade is providing a mechanism allowing rapid dissemination throughout the world, as well as distributing new resistances as they develop, which is probably happening at present with any new strain acquiring fumigant resistance. Pest control methods now take into account the problems associated with excessive pesticide usage and dependence to obtain control at an economic level.

There exist three very good reasons for reduction in pesticide use:

1. Development of resistance.
2. Cost of pesticides includes development, toxicology and registration
3. Environmental effects
   - destruction of beneficial species (pollinators such as bees)
   - irreversible effects on birds, fish and animals
   - long term effects on man himself
4. Misuse and abuse particularly in home garderners

The application of pesticides since the turn of the century can be classified under 2 broad categories:

1. Utilisation of marginally effective inorganic pesticides combined with labour intensive cultural control methods.

   This included some extremely deleterious compound such as the mercury based insecticides, arsenicals such as parts green, lead and calcium arsenate and the flouride insecticides. More recently, compounds of botanical origin such as nicotene, pyrethrum and rotenone, have 'had widespread application for field crop, orchard and domestic pest control.
2. Nearly sole reliance upon synthetic organic compounds applied on a calendar or preventative schedule.

DDT was the first of these insecticides, first synthesised by Zeidler in 1874, and recognised for its potenial in insect control in 1939 by Mueller against clothes moths. From the synthesis of DDT came a succession of related compounds (methoxychlor, dicofol) followed by the cyclodienes which are chlorinated cyclic hydrocarbons developed in 1945. Next came the organophosphates, carbonates, the synthetic pyrethroids to compounds attempting to mimic naturally occurring insect hormones applied at such a stage as to disrupt their normal life processes.

Both these categories in pesticide development have had associated and related problems such as:

1. Sub-economic control
2. Illegal chemical residues
3. Environmental contamination
4. Pesticide resistance
5. Subsequent resurgence of non-target secondary pests
6. Destruction of beneficial predators

Attempts have been made recently to resolve the contradictory demands of society that on the one hand asks for a constant supply o high quality food free of pest damage, but on the other hand is very much concerned with the potential negative side effects of pesticide usage. This approach to pest control has been referred to as integrated pest management (IPM).

The objectives of the system are to utilise an array of suitable control techniques rather than relying on a single disruptive tactic. It is important to realise that field applications of IPM allow sub-economic pest densities rather than demanding total eradication as the criterion for successful control; for the utilisation of biological control agents as a viable tactic depends on residual pest levels to insure the survival of natural enemies.

This is a practical approach which has proved to be an effective and economical control method in growing crop situations with minimum use of pesticides. But the situation is far more dramatic in storage where even the presence of beneficial parasites and predators are not tolerated. The growing crop can also absorb insect damage to a limited extent without any undue exonomic loss, but once a kernel of grain has been attacked, its nutritional value or potential for regeneration is lost forever.

10

Storage Losses and their Estimation

Kind of Storage Loss

Storage is one part of the post-harvest system through which food material passes on its way from field to consumer. Losses occurring in this post--harvest system, are finite and unlike growing crop losses they cannot be made up by further plant growth. Losses occuring during storage are affected by conditions prevailing in the pre-storage stages. Similarly, post-storage losses may be affected by conditions during storage.

There have been many attempts to define the term "loss" and much confusion has arisen since "loss" has frequently been used synonymously with "damage". Loss is a measurable decrease of the foodstuff which may be quantitative or qualitative. It should not be confused with damage which generally refers to the superficial evidence of deterioration, e.g. broken grains. Loss precludes utilisation, damage inhibits utilisation.

Loss may be considered in terms of either quantity or quality.

Quantitative loss is a physical loss of substance as shown by a reduction in weight or volume. It is the form of loss that can most readily be measured and valued.

Qualitative loss is more difficult to assess and is perhaps best identified through comparison with well defined standards. Nutritional loss and loss of seed are both aspects of quality losses.

Weight loss

Reduction in weight is obvious but it does not necessarily indicate loss. It may be due to reduced moisture content and this is recognised by a shrinkage factor. Weight loss results from the feeding of insects, rodents and birds or from spillage, due to improper handling or by the activities of pests.

Moisture changes may lead to an increase in weight and in some cases production of water by and insect infestation may partly offset the weight loss. In many instances weight loss may go undetected as the trader sells by volume. In commercial storage weight is the important factor sometimes leading to malpractices such as adulteration with water, stones, and earth to make up the deficiency. Usually some allowance is made for slight changes in moisture and also standard weight packages may be used.

Loss in Quality

Generally quality is assessed and products graded on the basis of appearance, shape, size, etc., but smell and flavour are sometimes included. Foreign matter content and contaminants are factors in loss of quality. Foreign matter may be in the form of insect fragments, grass, rodent hairs and excrete; weed seeds, parts of plants, earth, stones, glass, etc. Contaminants that cannot be readily removed, include soluble excretions of pests, oils, pesticides, pathogenic organisms spread by rodents, and toxins arising from fungal infections. Chemical changes may also be important, e.g. in oilseeds. Infestation in groundnuts may cause an increase in the free fatty acid level leading to rancidity in the oil, similarly in maize meal.

Nutritional Loss

This, is the product of both the quantitative and qualitative losses. Weight loss during storage (not due to a loss of moisture) is a measure of food loss but the latter may be proportionately larger owing to selective feeding by the pests. Rodents and moth larvae may preferentially attack the germ of the grain thus removing a large percentage of the protein and vitamin content, whereas weevils feeding mainly on the endosperm will reduce the carbohydrate content. Many pests may eat the bran of cereals reducing vitamins such as thiamin. Other storage factors such as moisture and fungal infection also lead to changes in vitamin content. In beans in particular, loss of protein is very important where there is infestation, as up to 25% of the dry matter may be crude protein.

Seed Viability Loss

This relates to loss in seed germination. Seed grain is usually more carefully stored owing to its greater potential value. Loss may be caused by changes of light, temperature, moisture, excessive respiration, infestation and, in some cases, the methods used to control infestation. Insects that selectively attack the germ will cause a greater loss in germination than others.

Commercial Losses

Commercial losses may be a consequence of any of the foregoing factors or be the preventive or remedial actions required, as well as equipment costs. These losses are generally incurred through a lack of knowledge, experience or managerial ability.

a) Monetary loss

Weight loss is an economic loss as is any downgrading of produce due to poor quality. Any control measure that has to be employed to render or keep the commodity saleable can be counted as an economic loss and is perhaps the most easily accountable loss. Losses in packaging and the costs of repacking due to rodent and handling damage, repairs and stoppages in machinery, damage to the fabric of the store are all economic losses that can be the result of infestation.

b) Loss of goodwill

This is not directly accountable but nonetheless it is very important, especially with regard to rising quality standards. A control measure that may seem uneconomic at first but leads to better custom or at least retains custom, is better than no control that leads to losing custom. This is particularly so in exports where a reputation for high quality produce is valuable to a country's economy.

c) Loss due to legal action

This may include damages awarded due to impairment of health of humans and animals, expenses incurred by third persons due to infestation traceable to a particular shipment, and various actions due to contamination.

Assessment of Storage Losses

The extent of losses occurring after harvest and particularly during storage has been the subject of considerable speculation because of the difficulties of reaching accurate assessments of the amount being lost. It is generally accepted that assessment of storage loss is difficult, particularly in relation to the rural situation in developing countries and there is a need for assessing storage losses. The extent of loss is important but not all important; other factors, economic, sociological or political, need to be considered in deciding on the methods of reducing loss. Measurement of storage loss then calls for an integrated multidisciplinary approach. Any project estimating losses should include the following essentials:

Clearly Defined Objectives

Loss measurements must be undertaken with a positive aim in view - that of loss reduction. When remedial measures have been introduced subsequent measures of loss using the same technique will enable the effectiveness of the remedial measures to be determined.

Reproducible Methods

Methods used should be adequately recorded so that they may be used by future workers for similar types of estimation. Obvious sources of error should be noted as they occur. Techniques used should be simple and only require relatively simple and robust apparatus if this has to be used under field conditions with unskilled personnel.

Representativeness

This includes the choice of the area to be investigated, choice of stores within the area and the sampling of these stores. Many factors may influence the 'typical' nature of the area so there has to be a balance between the ideal and practical approach. Factors that influence storage losses inculude climate, tradition, and extension influence. For a full statistical survey of losses at the farm level, farmers should be placed in categories that influence their storage, e.g. subsistence or commercial growers, and selection of stores made randomly within these various categories. This is known as statified random sampling but there are other methods which may be of use for various situations and can be found in textbooks on sampling of populations. The danger lies in extrapolating the data recorded and using it wrongly to sometimes support preconceived ideas; so one must be aware of the inherent limitations in such data.

Basic Framework

All loss estimations should be built up from a survey, a sampling method and the analysis of the sample to determine the loss.

Survey

This enables one to examine the area under investigation and define any factors influencing storage so that the selection of stores accounts for these differences. Field testing of any questionnaires or sampling techniques should be done during the initial survey to remove any practical difficulties. During this stage the criteria of loss should be decided upon and once the stores have been chosen any back-ground data needed for the assessment of their losses obtained.

Sampling

This refers to removal of grain samples from the stores. If the store is to be visited regularly, then a large sample should be taken when the grain is placed in store to provide basline data with which subsequent samples will be compared. Sampling once a month until the store is empty, coupled with records of grain consumption and disposal will enable an accurate estimate of the losses over the season to be obtained.

When such an intense sampling system cannot be undertaken, at least three samples need to be taken: 1) at the time of storage; 2) approximately halfway through; and 3) about a month before the store is emptied. These samples should be taken from the whole store to determine the pattern of infestation. Note should be made of the way in which grain is removed and of the pattern of consumption. Sampling throughout the store causes grain to move and changes the pattern of infestation within the store so should not be used if a monthly sampling routine is operated. It is however necessary if samples are to be taken on very few occasions.

Analysis of the Sample

Testing seed grain for loss normally involves only a simple germination test using a set of sub-samples. Analysis for quality loss is best carried out using a local grading system so that valuation of quality deterioration may be made. This will involve such points as defective and damaged grain, extraneous material, etc.

Nutritional losses need specific analysis for the particular substance one is interested in and comparison with its concentration at storage. It is important to remember that a weight to weight comparison will give misleading results as the insect damaged grain would originally have weighed more. A volume to volume comparison will be better or a comparison of a fixed number of grains taken at random. Measurement of protein using nitrogen determination will be prone to error as insects contain and produce a considerable amount of non-protein nitrogen. It is also possible that when grains are cooked whole, more nutrients may be lost from insect damaged grain than from intact grain.

Weight losses in samples removed for analysis may be estimated in several ways but prior to these it is usual to sieve off the dust and insects. These would normally be removed under field conditions prior to consumption. The proportion of these sievings can be used as a component of weight loss and for quality loss. Identification of the insects present will help the choice of possible control measures and show the relative importance of the various species.

There are three main methods of measuring weight losses caused by insects:

a) *Volumetric:* These may also be referred to as bushel weight or bulk density methods. They are based on obtaining an accurate standard volume of grain using simple apparatus using fixed drop heights into a standardised volume measure.

b) *Gravimetric:* This method only involves weighing and counting. It needs smaller samples than volumetric techniques with a maximum of about 1,000 grains and is a useful, quick, field method.

c) *Indirect:* These methods use some factor related to weight loss which can easily be measured and then transformed into an estimate of loss using a graph or formula. A laboratory experiment is usually done first, to determine the relationship. The results are then applied to the field samples of the same variety infested by the same pest. The most commonly used factor is percentage damaged grain as it is easily measured. Unfortunately this is complicated in the larger grain by multiple infestation leading to several emergence holes per grain and variation in the weight loss per grain. This could be avoided by using number of emergence holes per 100 grains instead of percentage damaged grain.

Losses Caused by Fungi

A considerable proportion of the grain rejected by the farmer is often discarded because of mould and the presence of infected grain causes a drop in quality grading. Therefore the impact of fungal infection on loss can be estimated by including the separation of mould damage from other types of damage during the analysis.

Losses Caused by Vertebrate Pests

Vertebrates such as rodents and birds frequently remove whole grains from store so losses caused by them are often obtained by difference. Losses remaining unaccounted for are attributed to them. Some attempts to estimate losses have been made using population census and feeding trials. Both pests utilise stored food as part of their diet only so feeding trials may overestimate the consumption of stored food.

Dry Matter Loss

The quantity of the grain from the time of harvest until it reaches the ultimate consumers decreases due to natural causes that can be difficult to prevent. In the warehouse, the quality and quantity of the grains received may not always be the same as that issued after several days or months of storage. Physical loss of grain in storage is the result of pest infestation, microbial infection, change in moisture content, poor handling and grain respiration and microorganic/microbial consumption. A recognition and understanding of the factors that contributes to dry matter or quantitative loss may be helpful In formulating possible preventive measures against grain losses particularly in storage.

Grain and Microorganisms Respiration

Quality of the grain is usually associated with the dry matter or quantitative loss that occurs because of respiration. Teter further noted that the actual respiration or breathing of grain itself is small. However, various organisms that live on grain are always present and respond in reasonably predictable ways to the ambient environment that consists of:

1) humidity or grain moisture,
2) temperature,
3) presence of toxins such as acids,
4) presence of microorganisms that cause deterioration such as bacteria, yeasts, and molds, and
5) physical stresses such as processing blows, heat stress, moisture stress and physical losses that reduce the quantity of materials.

Moreover, the percentage of losses due to bacteria and fungi is connected to a considerable number of factors such as species of microorganisms, moisture content, temperature, aeration conditions, sanitation, degree of infestation, and duration of storage. The concept of loss due to respiration has not been readily accepted in the assessment of losses in storage due to difficulty in quantification. Recently, Teter suggested that the percent dry matter can be calculated using the equation:

$$\% \text{ DML/day} = (CO_2)\ 10^{-3}\ (0.682)$$

Carbon dioxide generation can be related to dry matter loss where, for each gram of CO_2 given off, 0.682 gram of dry matter is lost.

For rough rice, equations developed from the laboratory and field are:

$\log (CO_2) = 0.44\ Mw - 6.08$ For Mw (13.3 to 15.6%)

$\log (CO_2) = 0.21\ Mw - 3.04$ For Mw (10.0 to 13.2%)

$\log (CO_2) = 2.39 \sin 7\ 5\ (Mw - B)$ For Mw (15.7 to 28%)

where:

CO_2 = mg CO_2 per 100 g dry matter per day

Mw = % moisture, wet basis

The general equation for rough rice for the range 10 to 17% is:

$$\log (CO_2) = AMw - B$$

where A and B are constants as shown in Table 1.

Seib, et al formulated the equation for dry matter loss using grain moisture content, grain temperature and relative humidity, and length of storage. The equation which is rather empirical is as follows:

$$DML = 1 - Exp - (At)\ Exp\ D(T - 60)\ Exp\ E(W - 0.14)$$

where:

DML = decimal value of dry matter loss

ACDE = constants

t = time elapsed, hours/1000

T = temperature, °F

W = moisture content, wet basis, decimal from

The constants are:

Palay	A	C	D	E
Long grain	0.001889	0.7101	0.02740	31.63
Medium grain	0.000914	0.6540	0.03756	33.61

In the study on Warehouse Inventory conducted by NAPHIRE, the estimated loss due to respiration of rough rice and microorganisms was calculated using the formula:

LR = I ave. wt. of DM stored I I (0.68) (| 0 44 | | Mw - 11 08 |) x I storage period I

where:

DM = dry matter

Mw = average moisture content of the grain stored, wet basis

Teter summarises the following principles that should always be considered in the estimation of dry matter loss:

1. Deteriorating grain generates 10.9 times the heat required to evaporate the water generated.
2. Badly deteriorated grain has a higher percentage of protein than the same grain in a sound condition.
3. Water generated by grain deterioration must be removed.
4. Deterioration continues as a summation of all spoilages encountered in the history of the grain.
5. When grain begins to heat, move it.
6. Unsuitable climatic air with deterioration index of over 3.0 to 5.0 should not be used.
7. The deterioration at the end of the storage time should not exceed 0.8% dry matter loss.

Loss Caused by Insects

Based on the capacity to infest sound kernels, insects are classified as either primary or secondary. Primary insects are those that make initial attack on fresh grain while the latter are those that feed on grain after it has been initially bored. Damage done by these insects consists of contamination and direct grain loss. Caliboso stated that a rice weevil can eat 14 milligrams during its developmental period from egg to adult.

A researcher from Tropical Products Institute recommends a "weight per unit volume method of estimating weight loss of insect infested grain." On the other hand, researchers from CRIA Grain Processing Research Center, monitored loss and insect damage for shelled corn in four (4) different containers and formulated the equation based on the relationship between loss and insect damage.

$$\% \text{ dry master} = .1 + x^{(0.0085x+0.15)}$$

where x is % of insect damaged kernels.

In storage loss assessment, Quitco and Quindoza used the converted percentage damage method to obtain a rough indication of loss caused by insects.

$$L1 = I\ 0.5\ I \times I\ \%\ \text{damage} \times \text{dry wt.}\ 1$$

where:

$$\%\text{insect damage grain} = \frac{\text{No. of insect damaged grain}}{1{,}000 \text{ grains}} \times 100$$

Loss due to Rodents

In Philippine storage, rodents are considered a serious pest that causes considerable amount of loss. According to Rubio, the annual rodent damage was estimated at 0.8 to 4.12 cavans per ricemillwarehouse in Laguna. In Nueva Ecija, Aganon reported that the annual grain loss per warehouse due to rodent contamination and spillage was estimated at 1.92 to 2.93 cavans. Based on previous researches, the major species affecting food in Philippine storage are the Norway rat, rattus norvegicus, the common ricefield rat, Rattus rattus mindanensis, and the house mouse, Mus musculus. Benigno found that rodents weighing more than 50 grams consume 15%. Sayaboc, et al., revealed that rodents have preference for whole kernels, and that rough rice constitute 95% of their diet. Considerably more grains are lost through spillage and contamination.

Caliboso and Teter suggested that to assess losses due to rodent, a constant estimate of the population density is needed. This can be obtained through trapping or by rougher methods such as counting of trails. The Zippin's method was found useful in estimating population density in the warehouse.

Based on the average body weight of Mus spe. (161 kg.), Rattus mindanensis (160 kg.), and Rattus norvegicus (330 g.), the daily intake of food constitutes 95% of the total diet. Thus, the daily loss can be computed by the formula:

$$L_R = (Pd_1 \times di_1) + (Pd_2 \times Di_2) + (Pd_3 \times Di_3)$$

where:

Pd_1 = Population density of specie 1 (n)

Di_1 = Daily intake of species (g)

For example, in a particular warehouse, there are 10 Norway rats, 3 Philippine rice-field rats, and 2 mice, the loss can be calculated as:

$$L_R = (10 \times 0.0314)+(3 \times 0.0152)+(2 \times 0.0025) = 0.3646 \text{ kg./day}$$

Rodent infestations may be controlled by three general methods: biological, physical and chemical. The use of these methods depends upon the prevailing storage or warehouse conditions. However, in most of the National Food Authority (NFA) warehouses, rats were controlled through good structural design of storage facilities. Warehouse floors were elevated, and windows or airvents are screened to serve as physical barriers to exclude rodents. If properly maintained, these warehouses could remain rat-proofed for many years.

Loss due to Birds

In the Philippines, the species of bird pests which feed on crops such as rice and sorghum are known locally as "maya" or Philippine weavers. Losses caused by these birds in stored products and often observed but seldom quantified. Grain on floor due to careless handling attracts birds and regular spillage may lead to establishment of a resident population.

Initial results of NAPHIRE study indicate that stored grain weight losses attributed to birds may be calculated in the same way for rodents. A population estimate may be based on visual counts or trappings. The mean body weight of Passer montanus, a bird species that is predominant in Philippine grain stores, is approximately 20 9. The daily intake was estimated at 30% of the mean body weight. Paddy constitutes 91 % of the total daily consumption. Thus, a single bird may consume 0.0055 kg. per day.

If estimated bird population is 150, loss is determined by:

$$L_B = \text{(Population Density) (Daily intake, kg.)} = (150)(0.0055) = 0.825 \text{ kg./day}$$

Several methods of controlling bird infestations in grain stores were suggested such as chemical means (poison baiting), biological methods, and physical methods. Caliboso recommends that the probable best method of bird control is to keep them off from the warehouse by bird proofing with nets or screen wire on all possible entrances like air vents and windows. Utmost care should be taken to keep the store and surrounding area clear of spillage and attractive food debris.

11

International Code of Conduct on the Distribution and Use of Pesticides

The action by FAO to develop, in consultation with appropriate United Nations agencies and other organisations, an International Code of Conduct on the Distribution and Use of Pesticides follows and accompanies many other events, some going back 25 years. All these events were designed to benefit the international community and to serve to increase international confidence in the availability, regulation, marketing and use of pesticides for the improvement of agriculture, public health and personal comfort.

One of the basic functions of the Code, which is voluntary in nature, is to serve as a point of reference, particularly until such time as countries have established adequate regulatory infrastructures for pesticides.

The Director-General of FAO in 1981 suggested that such a Code could help to overcome a number of difficulties associated with pesticides. The FAO Panel of Experts on Pesticide Specifications, Registration Requirements and Application Standards, at its meeting in 1982, agreed that activities involving the export and import of pesticides, and thereby their safe use, might be best dealt with through the adoption of a Code of Conduct. To that end a working paper was prepared for the FAO Second Government consultation on International Harmonisation of Pesticide Registration Requirements, Rome, 11-15 October 1982. The formal decision to develop the Code was taken at that Consultation, which recommended that FAO, in consultation with the appropriate United Nations organisations and bodies and international organisations outside the United Nations system, should draft a Code (1). The Code itself was adopted by the FAO Conference at its Twenty-third Session in 1985 by way of Resolution 10185, which appears as an Annex to the present publication.

A number of governments and organisations have expressed concern about the propriety of supplying pesticides to countries which do not have infrastructures to register pesticides and thereby to ensure their safe and effective use. It should be noted that the development of national regulatory programmes is the first priority of FAO activities in this field. There has also been concern over the possibility that residues of certain pesticides, not needed or not permitted in particular countries, are present in imported agricultural commodities produced in other countries where the use of such pesticides is not restricted. While recognising that it is impossible to eliminate all such occurrences, because of diverging pest control needs, it is none the less essential that every effort be made to apply pesticides only in accordance with good and recognised practices. It is at the same time important for industrially developed countries to recognise, in their regulatory activities concerning residues, the pest control needs of developing countries, particularly the needs of countries in tropical regions.

In the absence of an effective pesticide registration process and of a governmental infrastructure for controlling the availability of pesticides, some countries importing must heavily rely on the pesticide industry to promote the safe and proper distribution and use of pesticides. In these circumstances foreign manufacturers, exporters and importers, as well as local formulators, distributors, repackers, advisers and users, must accept a share of the responsibility for safety and efficiency in distribution and use.

The role of the exporting country needs to be considered. Much emphasis has been given recently to the desirability of regulating the export of pesticides from producing countries. It is generally accepted that no company should trade in pesticides without a proper and thorough evaluation of the pesticide, including any risks. However, the fact that a product is not used or registered in a particular exporting country is not necessarily a valid reason for prohibiting the export of that pesticide. Developing countries are mostly situated in tropical and semi-tropical regions.

Their climatic, ecological, agronomic, social, economic and environmental conditions and therefore their pest problems are usually quite different from those prevailing in countries in which pesticides are manufactured and exported. The government of the exporting country, therefore, is in no position to judge the suitability, efficacy, safety or fate of the pesticide under the conditions in the country where it may ultimately be used. Such a judgement must, therefore, be made by the responsible authority in the importing country in consultation with industry and other government authorities in the light of the scientific evaluation that has been made and a detailed knowledge of the conditions prevailing in the country of proposed use.

The export to developing countries of pesticides which have been banned in one or more other countries or whose use has been severely restricted in some industrialised countries has been a subject of public concern which has led to intensive discussions on whether the exporting country should assume responsibility for the marketing and use of such products in the importing country. In this respect it is essential to note that when pesticides are banned, the reasons are toxicological, environmental or social. Valid and adequate toxicological reasons justifying banning a product are of concern, though not necessarily of equal importance, to most countries. Consequently, such products should not be exported or imported without careful consideration of the toxicological implications for those likely to be exposed.

While a Code of Conduct may not solve all problems, nevertheless it should go a long way toward defining and clarifying the responsibilities of the various parties involved in the development, distribution and use of pesticides, and it should be of particular value in countries which do not yet have control procedures. Where there is a pesticide regulatory process in a country, the need for a Code of Conduct will obviously be less than where there is no such scheme in operation.

The Code of Conduct is not a short or simple document, mainly because the nature, properties, uses and effects of pesticides are diverse and therefore require comprehensive consideration. Furthermore, the strong public pressure for banning or restricting the use of some effective and much needed pesticides often stems from a lack of understanding of the many important issues involved. This document is designed, therefore, also to provide the general public with some basic guidance on these issues.

Text of the Code

Article 1. Objectives of the Code

The objectives of this Code are to set forth responsibilities and establish voluntary standards of conduct for all public and private entities engaged in or affecting the distribution and use of pesticides, particularly where there is no or an inadequate national law to regulate pesticides.

The Code describes the shared responsibility of many segments of society, including governments, individually or in regional groupings, industry, trade and international institutions, to work together so that the benefits to be derived from the necessary and acceptable use of pesticides are achieved without significant adverse effects on people or the environment. To this end, all references in this Code to a government or governments shall be deemed to apply equally to regional groupings of governments for matters falling within their areas of competence.

The Code addresses the need for a cooperative effort between governments of exporting and importing countries to promote practices which ensure efficient and safe use while minimising health and environmental concerns due to improper handling or use.

The entities which are addressed by this Code include international organisations; governments of exporting and importing countries; industry, including manufacturers, trade associations, formulators and distributors; users; and public sector organisations such as environmental groups, consumer groups and trade unions.

The standards of conduct set forth by this Code:

- encourage responsible and generally accepted trade practices;
- assist countries which have not yet established controls designed to regulate the quality and suitability of pesticide products needed in that country and to address the safe handling and use of such products;
- promote practices which encourage the safe and efficient use of pesticides, including minimising adverse effects on humans and the environment and preventing accidental poisoning from improper handling;
- ensure that pesticides are used effectively for the improvement of agricultural production and of human, animal and plant health.

The Code is designed to be used, within the context of national law, as a basis whereby government authorities, pesticide manufacturers, those engaged in trade and any citisens concerned may jidge whether their proposed actions and the actions of others constitute acceptable practices.

Article 2. Definitions

For the purpose of this Code:

Active ingredient means the biologically active part of the pesticide present in a formulation.

Advertising means the promotion of the sale and use of pesticides by print and electronic media, signs, displays, gift, demonstration or word of mouth.

Banned means a pesticide for which all registered uses have been prohibited by final government regulatory action, or for which all requests for registration or equivalent action for all uses have, for health or environmental reasons, not been granted.

Common name means the name assigned to a pesticide active ingredient by the International Standards Organisation or adopted by national standards authorities to be used as a generic or nonproprietary name for that particular active ingredient only.

Distinguishing name means the name under which the pesticide is labelled, registered and promoted by the manufacturer and which, if protected under national legislation, can be used exclusively by the manufacturer to distinguish the product from other pesticides containing the same active ingredient.

Distribution means the process by which pesticides are supplied through trade channels on local or international markets.

Environment means surroundings, including water, air, soil and their interrelationship as well as all relationships between them and any living organisms.

Extension service means those entities in the country concerned responsible for the transfer of information and advice to farmers regarding the improvement of agricultural practices, including production, handling, storage and marketing.

Formulation means the combination of various ingredients designed to render the product useful and effective for the purpose claimed; the form of the pesticide as purchased by users.

Hazard means the likelihood that a pesticide will cause an adverse effect under the conditions in which it is used.

Integrated pest management means a pest management system that, in the context of the associated environment and the population dynamics of the pest species, utilises all suitable techniques and methods in as compatible a manner as possible and maintains the pest populations at levels below those causing economically unacceptable damage or loss.

Label means the written, printed or graphic matter on, or attached to, the pesticide; or the immediate container thereof and the outside container or wrapper of the retail package of the pesticide.

Manufacturer means a corporation or other entity in the public or private sector or any individual engaged in the business or function of manufacturing a pesticide active ingredient or preparing its formulation or product

Marketing means the overall process of product promotion, including advertising, product public relations and information services as well as distribution and selling on local or international markets.

Maximum residue limit (MRL) means the maximum concentration of a residue that is legally permitted or recognised as acceptable in or on a food, agricultural commodity or animal feedstuff.

Packaging means the container together with the protective wrapping used to carry pesticide products via wholesale or retail distribution to users.

Pesticide means any substance or mixture of substances intended for preventing, destroying or controlling any pest, including vectors of human or animal disease, unwanted species of plants or animals causing harm during or otherwise interfering with the production, processing, storage, transport, or marketing of food, agricultural commodities, wood and wood products or animal feedstuffs, or which may be administered to animals for the control of insects, arachnids or other pests in or on their bodies. The term includes substances intended for use as a plant-growth regulator, defoliant, desiccant, or agent for thinning fruit or preventing the premature fall of fruit, and substances applied to crops either before or after harvest to protect the commodity from deterioration during storage and transport.

Pesticide industry means all those organisations and individuals engaged in manufacturing, formulating or marketing pesticides and pesticide products.

Pesticide legislation means any laws or regulations introduced to regulate the manufacture, marketing, storage, labelling, packaging and use of pesticides in their qualitative, quantitative and environmental aspects.

Poison means a substance that can cause disturbance of structure or function, leading to injury or death when absorbed in relatively small amounts by human beings, plants or animals.

Poisoning means occurrence of damage or disturbance caused by a poison, and includes intoxication.

Product means the pesticide in the form in which it is packaged and sold; it usually contains an active ingredient plus adjuvants and may require dilution prior to use.

Protective clothing means any clothes, materials or devices thet are designed to provide protection from pesticides when they are handled or applied.

Public-sector groups means scientifice associations; farmer groups; citizens' organisations; environmental, consumer and health organisations; and labour unions.

Registration means the process whereby the responsible national government authority approves the sale and use of a pesticide following the evaluation of comprehensive scientific data demonstrating that the product is effective for the purposes intended and not unduly hazardous to human or animal health or the environment.

Repackaging means the transfer of pesticide from any commercial package into any other, usually smaller, container for subsequent sale.

Residue means any specified substances in food, agricultural commodities, or animal feed resulting from the use of a pesticide. The term includes any derivatives of a pesticide, such as conversion products, metabolites, reaction products, and impurities considered to be of toxicological significance. The term "pesticide residue" includes residues from unknown or unavoidable sources as well as known uses of the chemical.

Responsible authority means the government agency or agencies responsible for regulating the manufacture, distribution or use of pesticides and more generally for implementing pesticide legislation.

Risk means the expected frequency of undesirable effects of exposure to the pesticide.

Severely restricted - a limited ban - means a pesticide for which virtually all registered uses have been prohibited by final government regulatory action but certain specific registered use or uses remain authorised.

Toxicity means a physiological or biological property which determines the capacity of a chemical to do harm or produce injury to a living organism by other than mechanical means.

Trader means anyone engaged in trade, including export, import, formulation and domestic distribution.

Use pattern embodies the combination of all factors involved in the use of a pesticide, including the concentration of active ingredient in the preparation being applied, rate of application, time of treatment, number of treatments, use of adjuvants and methods and sites of application which determine the quantity applied, timing of treatment and interval before harvest, etc.

Article 3. Pesticide Management

Governments have the overall responsibility and should take the specific powers to regulate the distribution and use of pesticides in their countries.

The pesticide industry should adhere to the provisions of this Code as a standard for the manufacture, distribution and advertising of pesticides, particularly in countries lacking appropriate legislation and advisory services.

Governments of exporting countries should help to the extent possible, directly or through their pesticide industries, to:

- provide technical assistance to other countries, especially those with shortages of technical expertise, in the assessment of the relevant data on pesticides, including those provided by industry;
- ensure that good trading practices are followed in the export of pesticides, especially to those countries with no or limited regulatory schemes.

Manufacturers and traders should observe the following practices in pesticide management, especially in countries without legislation or means of implementing regulations:

- supply only pesticides of adequate quality, packaged and labelled as appropriate for each specific market;
- pay special attention to formulations, presentation, packaging and labelling in order to reduce hazard to users, to the maximum extent possible consistent with the effective functioning of the pesticide in the particular circumstances in which it is to be used;
- provide, with each package of pesticide, information and instructions in a form and language adequate to ensure safe and effective use

– retain an active interest in following their products to the ultimate consumer, keeping track of major uses and the occurrence of any problems arising in the actual use of their products as a basis for determining the need for changes in labelling, directions for use, packaging, formulation or product availability.

Pesticides whose handling and application require the use of uncomfortable and expensive protective clothing and equipment should be avoided, especially in the case of small-scale users in tropical climates.

National and international organisations, governments, and pesticide industries should take action in coordinated efforts to disseminate eductional materials of all types to pesticide users, farmers, farmers" organisations, agricultural workers, unions and other interested parties. Similarly, affected parties should seek and understand educational materials before using pesticides and should follow proper procedures.

Governments should allocate high priority and adequate resources to the task of effectively managing the availability, distribution and use of pesticides in their countries.

Concerted efforts should' be made by governments and pesticide industries to develop and promote integrated pest management systems and the use of safe, efficient, cost-effective application methods. Public-sector groups and international organisations should actively support such activities.

International organisations should provide information on specific pesticides and give guidance on methods of analysis through the provision of criteria documents, fact sheets, training sessions, etc.

It is recognised that the development of resistance of pests to pesticides can be a major problem. Therefore, governments, industry, national institutions, international organisations and public sector groups should collaborate in developing strategies which will prolong the useful life of valuable pesticides and reduce the adverse effects of the development of resistant species.

Article 4. Testing of Pesticides

Pesticide manufacturers are expected to:

- ensure that each pesticide and pesticide product is adequately and effectively tested by wellrecognised procedures and test methods so evaluate as to fully evaluate its safety, efficacy (2) and fate (3) with regard to the various anticipated conditions in regions or countries of use:
- ensure that such tests are conducted in accordance with sound scientific procedures and good laboratory practice (4)- the data produced by such tests, when evaluated by competent experts, must be capable of showing whether the product can be handled and used safely without unacceptable hazard to human health, plants, animals, wildlife and the environment (3);
- make available copies or summaries of the original reports of such tests for assessment by responsible government authorities in all countries where the pesticide is to be offered for sale. Evaluation of the data should be referred to qualified experts;
- take care to see that the proposed use pattern, label claims and directions, packages, technical literature and advertising truly reflect the outcome of these scientific tests and assessments;

- provide, at the request of a country, advice on methods for the analysis of any active ingredient of formulation that they manufacture, and provide the necessary analytical standards;
- provide advice and assistance for training technical staff in relevant analytical work. Formulators should actively support this effort;
- conduct residue trials prior to marketing in accourdance with FAO guidelines on good analytical practice and on crop residue data in order to provide a basis for establishing appropriate maximum residue limits (MRLs).

Each country should possess or have access to facilities to verify and exercise control over the quality of pesticides offered for sale, to establish the quantity of the active ingredient or ingredients and the suitability of their formulation.

International organisations and other interested bodies should, within available resources, consider assisting in the establishment of analytical laboratories in pesticide-importing countries, either on a country or on a multilateral regional basis; these laboratories should be capable of carrying out product and residue analysis and should have adequate supplies of analytical standards, solvents and reagents.

Exporting governments and international organisations must play an active role in assisting developing countries in training personnel in the interpretation and evaluation of test data.

Industry and governments should collaborate in conducting post-registration surveillance or monitoring studies to determine the fate and environmental effect of pesticides under field conditions (3).

Article 5. Reducing Health Hazards

Governments which have not already done so should:

- Implement a pesticide registration and control scheme along the lines set out in Article 6;
- Decide, and from time to time review, the pesticides to be marketed in their country, their acceptable uses and their availability to each segment of the public;
- Provide guidance and instructions for the treatment of suspected pesticide poisoning for their basic health workers, physicians and hospital staff;
- Establish national or regional poisoning information and control centres at strategic locations to provide immediate guidance on first aid and medical treatment, accessible at all times by telephone or radio. Governments should collect reliable information about the health aspects of pesticides. Suitably trained people with adequate resources must be made available to ensure that accurate information is collected;
- keep extension and advisory services, as well as farmers' organisations, adequately informed about the range of pesticide products available for use in each area;
- ensure, with the cooperation of industry, that where pesticides are available through outlets which also deal in food, medicines, other products for internal consumption or topical application, or clothing, they are physically segregated from other merchandise, so as to avoid any possibility of contamination or of mistaken identity. Where appropriate, they should be clearly marked as hazardous materials. Every effort should be made

to publicise the dangers of storing foodstuffs and pesticides together.

Even where a control scheme is in operation, industry should:

- cooperate in the periodic reassessment of the pesticides which are marketed and in providing the poison control centres and other medical practitioners with information about hazards;
- make every reasonable effort to reduce hazard by:
  - making less toxic formulations available;
  - introducing products in readyto-use packages and otherwise developing safer and more efficient methods of application;
  - using containers that are not attractive ror subsequent reuse and promoting programmes to discourage their reuse;
  - using containers that are safe (e.g. not attractive to or easily opened by children), particularly for the more toxic homeuse products;
  - using clear and concise labelling;
- halt sale, and recall products; when safe use does not seem possible under any use directions or restrictions.

Government and industry should further reduce hazards by making provision for safe storage and disposal of pesticides and containers at both warehouse and farm level, and through proper siting and control of wastes from formulating plants.

To avoid unjustified confusion and alarm among the public, public-sector groups should consider all available facts and try to distinguish between major differences in levels of risk among pesticides and uses.

In establishing production facilities in developing countries, manufacturers and governments shoud cooperate to:

- adopt engineering standards and safe operating practices appropriate to the nature of the manufacturing operations and the hazards involved;
- take all necessary precautions to protect the health and safety of operatives, bystanders and the environment;
- maintain quality-assurance procedures to ensure that the products manufactured comply to the relevant standards of purity, performance, stability and safety.

Article 6. Regulatory and technical requirements

Governments should

- take action to introduce the necessary legislation for the regulation, including registration, of pesticides and make provisions for its effective enforcement, including the establishment of appropriate educational, advisory, extension and health care services; the FAO guidelines for the registration and control of pesticides (9) should be followed, as far as possible, taking full accuount of local needs, social and economic conditions, levels of literacy, climatic conditions and availability of pesticide application equipment;
- strive to establish pesticide registration schemes and infrastructures under which products can be registered prior to domestic use and, accordingly, ensure that each pesticide product is registered

under the laws or regulations of the country of use before it can be made available there;

- protect the proprietary rights to use of data;
- collect and record data on the actual import, formulation and use of pesticides in each country In order to assess the extent of any possible effects on human health or the environment, and to follow trends in use levels for economic and other purposes.

The pesticides industry should:

- provide an objective appraisal together with the necessary supporting data on each product;
- ensure that the active ingredient and other ingredients of pesticide preparations marketed correspond in identity, quality, purity and composition to the substances tested, evaluated and cleared for toxicological and environmental acceptability;
- ensure that active ingredients and formulated products for pesticides for which international specifications have heen developed conform with the specifications of FAO (8), where intended for use in agriculture; and with WHO pesticide specifications (10), where intended for use in public health;
- verify the quality and purity of the pesticides offered for sale;
- when problems occur, voluntarily take corrective action, and when requested by governments, help find solutions to difficulties.

Article 7. Availability and Use

Responsible authorities should give special attention to drafting rules and regulations on the availability of pesticides. These should be compatible with existing levels of training and expertise in handling pesticides on the part of the intended users. The parameters on which such decisions are based vary widely and must be left to the discretion of each government, bearing in mind the situation prevailing in the country.

In addition, governments should take note of and, where appropriate, follow the WHO classifications of pesticides by hazard (II) and associate the hazard class with well-recognised hazard symbols as the basis for their own regulatory measures. In any event, the type of formulation and method of application should be taken into account in determining the risk and degree of restriction appropriate to the product.

Two methods of restricting availability can be exercised by the responsible authority: not registering a product; or, as a condition of registration, restricting the availability to certain groups of users in accordance with national assessments of hazards involved in the use of the product in the particular country.

All pesticides made available to the general public should be packaged and labelled in a manner which is consistent with the FAO guidelines on packaging (12) and labelling (13) and with appropriate national regulations.

Prohibition of the importation, sale and purchase of an extremely toxic product may be desirable if control measures or good marketing practices are insufficient to ensure that the product can be used safely. However, this is a matter for decision in the light of national circumstances.

Article 8. Distribution and Trade

Industry should:

- test all pesticide products to evaluate safety with regard to human health and the environment prior to marketing, as provided in Article 4, and ensure that all pesticide products are likewise adequately tested for efficacy and stability and crop tolerance, under procedures that will predict performance under the conditions prevailing in the region where the product is to be used, before they are offered there for sale;
- submit the results of all such tests to the local responsible authority for independent evaluation and approval before the products enter trade channels in that country;
- take all necessary steps to ensure that pesticides entering international trade conform to relevant FAO, (8), WHO (10) or equivalent specifications for composition and quality (where such specifications have been developed) and to the principles embodied in pertinent FAO guidelines, and in rules and regulations on classification and packaging, marketing, labelling and documentation laid down by international organisations concerned with modes of transport;
- undertake to see that pesticides which are manufactured for export are subject to the same quality requirements and standards as those applied by the manufacturer to comparable domestic products;
- ensure that pesticides manufactured or formulated by a subsidiary company meet appropriate quality

requirements and standards which should be consistent with the requirements of the host country and of the parent company;

- encourage importing agencies, national or regional formulators, and their respective trade organisations to cooperate in order to achieve fair practices and safe marketing and distribution practices and to collaborate with authorities in stamping out any malpractices within the industry;
- recognise that the recall of a pesticide by a manufacturer and distributor may be desirable when faced with a pesticide which represents an unacceptable hazard to human and animal health and the environment when used as recommended, and cooperate accordingly;
- endeavour to ensure that pesticides are traded by and purchased from reputable traders, who should preferably be members of a recognised trade organisation;
- see that persons involved in the sale of any pesticide are trained adequately to ensure that they are capable of providing the buyer with advice on safe and efficient use;
- provide a range of pack sizes and types which are appropriate for the needs of small-scale farmers and other local users to avoid handling hazards and the risk that resellers will repackage products into unlabelled or inappropriate containers.

Governments and responsible authorities should take the necessary regulatory measures to prohibit the repackaging, decanting or dispensing of any pesticide in food or beverage containers and should rigidly enforce punitive measures that effectively deter such practices.

Governments of countries importing food an agricultural commodities should recognise good agricultural practices in countries with which they trade and, in accordance with recommendations of the Codex Alimentarius Commission, should establish a legal basis for the acceptance of pesticide residues resulting from such good agricultural practice.

Article 9. Information Exchange

The government of a pesticide-exporting country which takes action to ban or severely restrict the use or handling of a pesticide in order to protect health or the environment domestically should notify, directly or indirectly, the designated national authorities in other countries of the action it has taken.

The purpose of the notification regarding control action is to give competent authorities in other countries the opportunity to assess the risks associated with the pesticide, and to make timely and informed decisions as to the importation and use of the pesticides concerned, after taking into account local, public-health, economic, environmental and administrative conditions. The minimum information to be provided for this purpose should be:

- the identity (common name, distinguishing name and chemical name);
- a summary of the control action taken and of the reasons for it - if the control action bans or restricts certain uses but allows other uses, such information should be included;
- the fact that additional information is availabe, and the name and address of the contact point in the country of export to which a request for further information should be addressed.

If export of a banned or severely restricted pesticide occurs, the country of export should ensure that necessary steps are taken to provide the designated national authority of the country of import with relevant information.

The purpose of information regarding exports is to remind the country of import of the original notification regarding control action and to alert it to the fact that an export is expected or is about to occur. The minimum information to be provided for this purpose should be:

- a copy of, or reference to, the information provided at the time of the notification of control action;
- indication that an export of the chemical concerned is expected or is about to occure.

Notification of control action should be provided as soon as practicable after the control action is taken. For pesticides banned or severely restricted before the implementation of the Code, an inventory of prior control action should be provided to the International Register of Potentially Toxic Chemicals (IRPTC), unless such information has already been provided.

Provision of information regarding exports should take place at the time of the first export following the control action, and should recur in the case of any significant development of new information or condition surrounding the control action. It is the intention that the infomation should be provided prior to export.

The provision of such information by the exporting country must take into account protection of the confidentiality of data in the importing country.

Governments of importing countries should:

- establish internal procedures for the receipt and handling of such information from the exporting country;

- ensure that such information received is not used in any manner which would be inconsistent with the provisions of the General Agreement on Tariffs and Trade (GATT).

Article 10. Labelling, Packaging, Storage and Disposal

All pesticide containers should be clearly labelled in accordance with applicable international guidelines, such as the FAO guidelines on good labelling practice.

Industry should use labels that:

- include recommendations consistent with those of the recognised research and advisory agencies in the country of sale;
- include appropriate symbols and pictograms whenever possible, in addition to written instructions, warnings and precautions;
- in international trade, clearly show appropriate WHO hazard classification of the contents (11) or, if this is inappropriate or inconsistent with national regulations, use the relevant classification;
- include, in the appropriate language or languages, a warning against the reuse of containers, and instructions for the safe disposal or decontamination of empty containers;
- identify each lot or batch of the product in numbers or letters that can be read, transcribed and communicated by anyone without the need for codes or other means of deciphering;
- are marked with the date (month and year) of formulation of the lot or batch and with relevant information on the storage stability of the product.

Industry should ensure that:

- packaging, storage and disposal of pesticides

conform in principle to the FAO guidelines for packaging and storage, the FAO guidelines for the disposal of waste pesticides and containers, and WHO specifications for pesticides used in public health;

- in cooperation with governments, packaging or repackaging is carried out only on licensed premises where the responsible authority is convinced that staff are adequately protected against toxic hazares, that the resulting product will be properly packaged and labelled, and that the content will conform to the relevant quality standards.

Governments should take the necessary regulatory measures to prohibit the repacking, decanting or dispensing of any pesticide into food or beverage containers in trade channels and rigidly enforce punitive measures that effectively deter such practices.

Article 11. Advertising

Industry should ensure that:

- all statements used in advertising are capable of technical substantiation;
- advertisements do not contain any statement or visual presentation which, directly or by implication, omission, ambiguity or exaggerated claim, is likely to mislead the buyer, in particular with regard to the safety of the product, its nature, composition, or suitablility for use, or official recognition or approval;
- pesticides which are legally restricted to use by registered operators are not publicly advertised through journals other than those catering for such

operations, unless the restricted availability is clearly and prominently shown;

- no firm or individual in any one country simultaneously markets different pesticide active ingredients or combinations of ingredients under a single distinguishing name;
- advertising does not encourage uses other than those specified on the approvel label;
- promotional material does not include use recommendations at variance with those of the recognised research and advisory agencies;
- advertisements do not misuse research results or quotations from technical and scientific literature; and scientific jargon and irrelevances are not used to make claims appear to have a scientific basis they do not possess;
- claims as to safety, including statements such as "safe", "non-poisonous", "harmless", "nontoxic", are not made, with or without a qualifying phrase such as "when used as directed";
- statements comparing the safety of different products are not made;
- misleading statements are not made concerning the effectiveness of the product;
- no guarantees or implied guarantees e.g. "more profits with...", "guarantees high yields" are given unless definite evidence to substantiate such claims is available;
- advertisements do not contain any visual representation of potentially dangerous practices, such as mixing or application without sufficient protective clothing, use near food, or use by or near children;

- advertising or promotional material draws attention to the appropriate warning phrases and symbols as laid down in the labelling guidelines;
- technical literature provides adequate information on correct practices, including the observance of recommended rates, frequency of applications, and safe pre-harvest intervals;
- false or misleading comparisons with other pesticides are not made;
- all staff involved in sales promotion are adequately trained and possess sufficient technical knowledge to present complete, accurate and valid information on the products sold;
- advertisements encourage purchasers and users to read the label carefully, or have label read to them if they cannot read.

International organisations and public sector groups should call attention to departures from this Article.

Governments are encouraged to work with manufacturers to take advantage of their marketing skills and infrastructure, in order to provide publicservice advertising regarding the safe and effective use of pesticides. This advertising could focus on such factors as proper maintenance and use of equipment, special precautions for children and pregnant women, the danger of reusing containers, and the importance of following label directions.

Article 12. Monitoring the Observance of the Code

The Code should be published and should be observed through collaborative action on the part of governments, individually or in regional groups, appropriate organisations and bodies of the United Nations system, international governmental organisations and the pesticide industry.

The code should be brought to the attention of all concerned in the manufacture, marketing and use of pesticides and in the control of such activities, so that governments, individually or in regional groupings, industry and international institutions understand their shared responsibilities in working together to ensure that the objectives of the Code are achieved.

All parties addressed by this Code should observe this Code and should promote the principles and ethics expressed by the Code, irrespective of other parties' ability to observe the Code. The pesticide industry should cooperate fully in the observance of the Code and promote the principles and ethics expressed by the Code, irrespective of a government's ability to observe the Code.

Independently of any measures taken with respect to the observance of this Code, all relevant legal rules, whether legislative, administrative, judicial or customary, dealing with liability, consumer protection, conservation, pollution control and other related subjects should be strictly applied.

FAO and other competent international organisations should give full support to the observance of the Code, as adopted.

Governments should monitor the observance of the Code and report on progress made to the Director-General of FAO.

Governing Bodies should periodically review the relevance and effectiveness of the Code. The Code should be considered a dynamic text which must be brought up to date as required, taking into account technical, economic and social progress.

Annexure

FAO Conference Resolution 10/85: International Code of Conduct on the Distribution and Use of Pesticides

The Conference

Recognising that increased food production is a high priority need in many parts of the world and that this need cannot be met without the use of indispensable agricultural inputs such as pesticides,

Noting that FAO's study entitled Agriculture: toward 2000 foresees a steady increase in the worldwide use of pesticides,

Convinced that such growth in pesticide use is likely to take place in spite of necessary intensive parallel efforts to introduce biological and integrated pest control systems,

Acknowledging that pesticides can be hazardous to humans and the environment and that immediate action must be taken by all concerned, including governments, manufacturers, traders and users, to eliminate, as far as possible and within the scope of their responsibility, unreasonable risks, not only in the country of origin but also in the countries to which pesticides may be exported,

Being aware that the requirements for safe and proper use of pesticides in some developed countries have led to the adoption of complex systems of regulations and of enforcement mechanisms, but that many other countries have neither such mechanisms nor the necessary legislation, regulations or infrastructures to control the import, sale or use of pesticides,

Convinced that additional efforts are needed to enable such countries to control pesticides more effectively and to assess the hazards which could result from their use or misuse,

Recognising that a voluntary International Code of Conduct, based on internationally agreed technical guidelines, would provide a practical framework for the control of pesticides, especially in countries that do not have adequate pesticide registration and control schemes,

Noting that such a draft Code was reviewed by the Committee on Agriculture at its Eighth Session, and endorsed by the Council at its Eighty-eighth Session,

Having further noted the conclusions and recommendations of these bodies,

1. Hereby adopts a voluntary International Code of Conduct on the Distribution and Use of Pesticides as given in the annex to this Resolution;
2. Recommends that all FAO Member Nations promote the use of this Code in the interests of safer and more efficient use of pesticides and of increased food production;
3. Requests governments to monitor the observance of the Code, in collaboration with the Director-General who will report periodically to the Committee on agriculture;
4. Invites other United Nations agencies and other international organisations to collaborate in this endeavour within their respective spheres of competence.

Bibliography

Beane, K., "Managing Pest Birds on Buildings," *Common Sense Pest Control, VIII (3),* Summer, 1992.

Bennett, G.W. and J.M. Owens (ed)., *Advances in Urban Pest Management,* Van Nostrand and Reinhold, 1986.

Bottrell, D., *Integrated Pest Management,* Council on Environmental Quality, 1979.

Ebeling, W., *Urban Entomology,* University of California, Division of Agricultural Sciences, 1978.

Flint, M.L. and R. van den Bosch, *Introduction to Integrated Pest Management,* Plenum Press, 1981.

______________, "A Source Book on Integrated Pest Management", International Center for Integrated and Biological Control, University of California, 1977.

Haverty, M.I., "Termites", Pest Control, May, pp. 12-17 and 46-49, 1976.

Imholte, T.J., "Engineering for Food Safety and Sanitation", Technical Institute of Food Safety, 1984.

Lifton, B., "Bug Busters: Getting Rid of Household Pests without Dangerous Chemicals", McGraw-Hill Book Company, 1985.

Mallis, A., *Handbook of Pest Control: The Behavior, Life History, and Control of Household Pests,* 7th edition, Franzak & Foster, 1990.

Olkowski, W., H. Olkowski and S. Daar, "What is IPM?", *Common Sense Pest Control Quarterly,* Vol. 4, no. 3, Summer, 1988.

_______________, "Integrated Pest Management for the House Mouse", Bio-Integral Resource Center, Berkeley California, 1984.

Quarles, W., "Diatomaceous Earth for Pest Control," *The IPM Practitioner*, Vol. XIV, No. 5/6, May/June, 1992.

Truman, L.C., G.W. Bennet and W.L. Butts, *Scientific Guide to Pest Control Operations*, 4th edition, Harcourt Brace Janovich, 1988.

Ware, G., *Complete Guide to Pest Control-With and Without Chemicals*, Thomson Publications, Fresno, California,1980.

Index